I0065010

Weed and Pest Control

Weed and Pest Control

Owen Morales

R CALLISTO
REFERENCE

www.callistoreference.com

Callisto Reference,
118-35 Queens Blvd., Suite 400,
Forest Hills, NY 11375, USA

Visit us on the World Wide Web at:
www.callistoreference.com

© Callisto Reference, 2022

This book contains information obtained from authentic and highly regarded sources. All chapters are published with permission under the Creative Commons Attribution Share Alike License or equivalent. A wide variety of references are listed. Permissions and sources are indicated; for detailed attributions, please refer to the permissions page. Reasonable efforts have been made to publish reliable data and information, but the authors, editors and publisher cannot assume any responsibility for the validity of all materials or the consequences of their use.

ISBN: 978-1-64116-521-1 (Hardback)

Trademark Notice: Registered trademark of products or corporate names are used only for explanation and identification without intent to infringe.

Cataloging-in-Publication Data

Weed and pest control / Owen Morales.
    p. cm.
Includes bibliographical references and index.
ISBN 978-1-64116-521-1
1. Weeds--Control. 2. Agricultural pests--Control. 3. Pests--Control.
4. Plants, Protection of. I. Morales, Owen.
SB611 .W44 2022
632.5--dc23

Table of Contents

This book has been written, keeping in view that students want more practical information. Thus, my aim has been to make it as comprehensive as possible for the readers. I would like to extend my thanks to my family and co-workers for their knowledge, support and encouragement all along.

Weed control aims to stop noxious weeds from competing with the desired flora and fauna. It plays a vital role in agriculture. They also contaminate harvests and interfere with the management of desirable plants. The diverse methods of weed control are broadly divided into physical methods, cultural methods and chemical methods. The physical methods include covering the weed, manual removal, thermal methods, tillage and seed targeting. Some of the cultural methods of weed control are stale seed beds and crop rotation. The management of pests which have adverse effects upon human activities is termed as pest control. Biological pest control and pesticides are a few methods of controlling pests. Biological pest control relies on predation and usage of other organisms to control pests. This book brings forth some of the most innovative concepts and elucidates the unexplored aspects of weed and pest control. Some of the diverse topics covered in this book address the varied methods used within these fields. Those in search of information to further their knowledge will be greatly assisted by this book.

A brief description of the chapters is provided below for further understanding:

Chapter – Weed and its Types

Weed refers to the unwanted wild plants that grow in gardens and fields. It includes forest weeds, aquatic weeds, parasitic weeds, etc. These are also used for repelling pests, soil fertilization, as flavor enhancements and natural herbicide. This chapter has been carefully written to provide an easy understanding of these types of weeds and their uses.

Chapter – Types of Pests

Pest includes crops, livestock and forestry that impacts the human activities adversely. Ants, bees, fleas, myriapods, rodents, flies, bed bugs, wasps, etc. are some of its types. The topics elaborated in this chapter will help in gaining a better perspective about pests and their types.

Chapter – Weed Control: A Comprehensive Study

Weed control is the method which controls and manages the growth of noxious and invasive weeds. Mechanical weed control, cultural weed control, biological weed control, chemical weed control, integrated weed management, etc. fall under its domain. This chapter delves into these methods of weed control to provide an in-depth understanding of the subject.

Chapter – Pest Control Methods

Pest control is the management and regulation of pests which are harmful to humans. It can be categorized into biological, physical, mechanical and chemical methods of pest control including fungicides, trap crop, rodenticides, miticides, etc. This chapter discusses these methods of pest control in detail.

Chapter – Urban Pest Control

Urban pests refer to the parasitic microorganisms which effect the human health and damage wooden support structures. It includes mosquito control, bed bugs control, rodent control, etc. This chapter closely examines these urban pest control methods to provide an extensive understanding of the subject.

Owen Morales

1
Weed and its Types

Weed refers to the unwanted wild plants that grow in gardens and fields. It includes forest weeds, aquatic weeds, parasitic weeds, etc. These are also used for repelling pests, soil fertilization, as flavor enhancements and natural herbicide. This chapter has been carefully written to provide an easy understanding of these types of weeds and their uses.

Weed

Weed is a general term for any plant growing where it is not wanted. Ever since humans first attempted the cultivation of plants, they have had to fight the invasion by weeds into areas chosen for crops. Some unwanted plants later were found to have virtues not originally suspected and so were removed from the category of weeds and taken under cultivation. Other cultivated plants, when transplanted to new climates, escaped cultivation and became weeds or invasive species. The category of weeds thus is ever changing, and the term is a relative one.

Weeds interfere with a variety of human activities, and many methods have been developed to suppress or eliminate them. These methods vary with the nature of the weed itself, the means at hand for disposal, and the relation of the method to the environment. Usually for financial and ecological reasons, methods used on a golf course or a public park cannot be applied on rangeland or in the forest. Herbicide chemicals sprayed on a roadside to eliminate unsightly weeds that constitute a fire or traffic hazard are not proper for use on cropland. Mulching, which is used to suppress weeds in a home garden, is not feasible on large farms. Weed control, in any event, has become a highly specialized activity. Universities and agricultural colleges teach courses in weed control, and industry provides the necessary technology. In agriculture, weed control is essential for maintaining high levels of crop production.

The many reasons for controlling weeds become more complex with the increasing development of technology. Plants become weeds as a function of time and place. Tall weeds on roadsides presumably were not problematic prior to the invention of the automobile.

However, with cars and increasing numbers of drivers on roads, tall weeds became dangerous, potentially obscuring drivers' visibility, particularly at intersections. Sharp-edged grasses are nominal nuisances in a cow pasture; when the area is converted to a golf course or a public park, they become an actual nuisance. Poison oak (Toxicodendron diversilobum) is rather a pleasant shrub on a sunny hillside in the open country; in a camp ground it is a definite health hazard. Such examples could be given ad infinitum to cover every aspect of agriculture, forestry, highway, waterway and public land management, arboretum, park and golf-course care, and home landscape maintenance.

Weeds compete with crop plants for water, light, and nutrients. Weeds of rangelands and pastures may be unpalatable to animals, or even poisonous; they may cause injuries, as with lodging of foxtails (Alopecurus species) in horses' mouths; they may lower values of animal products, as in the cases of cockleburs (Xanthium species) in wool; they may add to the burden of animal care, as when horses graze in sticky tarweeds (Madia species). Many weeds are hosts of plant disease organisms. Examples are prickly lettuce (Lactuca scariola) and sow thistle (Sonchus species) that serve as hosts for downy mildew; wild mustards (Brassica species) that host clubroot of cabbage; and saltbrush (Atriplex species) and Russian thistle, in which curly top virus overwinters, to be carried to sugar beets by leafhoppers. Many weeds are hosts of insect pests, and a number are invasive species.

There are generally three types of common weed plants in regards to their growing characteristics. These include:

Annual types – Annual weeds germinate and spread by seed, having an average lifespan of one year. These include both winter and summer types. Winter annuals, like chickweed, germinate in late summer/early fall, go dormant in winter and actively grow during spring. Summer annuals, such as lambsquarters, germinate in spring, grow throughout summer and are gone with the arrival of cold weather.

- Biennial types: Biennial weeds complete their life cycle in two years, germinating and forming rosettes their first year and producing flowers and seeds their second year. Examples of these types include: bull thistle and garlic mustard.

- Perennial types: Perennial weeds return every year and normally produce long tap roots in addition to seeds. These weeds, which include dandelions, plantain, and purple loosestrife, are the most difficult to control.

In addition to their growing type, common weed plants may belong to one of two families: broadleaf (Dicot) or narrow leaf (Monocot). Broadleaf types have larger leaves and grow from tap roots or fibrous root systems, whereas narrow leaf or grasses have long narrow leaves and fibrous roots systems.

Weed info and Control

There are a number of weed control methods, depending on the weed and the gardener. Here are your options.

Cultural weed control: One of the easiest ways to control weeds is through prevention or cultural control. Close planting in the garden can reduce weed growth by eliminating open space. Cover crops are good for this as well. Adding mulch will prevent light from getting to weed seeds and prevents growth.

Mechanical weed control: Mechanical control of common weed plants can be accomplished through hand pulling, hoeing, digging or mowing (which slows growth and reduces seed formation). While these methods are effective, they can be time consuming.

Chemical weed control: Since many weeds, like dodder, ivy and kudzu, can become aggressive to the point of taking over, chemical control is sometimes necessary, and used normally a last resort. There are numerous herbicides available to help eliminate common weed plants.

Natural weed control: Generally, invasive weeds are well worth the trouble of removal. However, some weeds can actually be quite attractive in the garden, so why not consider allowing them to stay. This more natural weed control method results in a lush native environment when given their own designated spot. Some of these 'good weeds' include:

- Joe-pye weed: Tall stems of vanilla-scented rose-colored flower clusters.

- Chicory: Brilliant blue flowers.

- Hawkweed: Daisy-like blooms on fuzzy stems.

- Queen Anne's-lace: Lacy white, umbrella-shaped flower heads.

Forest Weeds

Competitive division of weeds in forestry is often made according to the degree of harmfulness of weeds to the trees. According to Vajda weeds in forestry are classified as either useful or harmful; Konstantinovic categorizes weeds into useful, harmful, or indifferent. According to this classification:

- Harmful weeds are plants that hinder tree development, and form thick cover.

- Indifferent weeds are plants that grow individually, form weak coverage and do not hinder development of cultivated plants.

- Useful weeds are plants with medical properties and plants that form fruits.

The role of light has been of particular importance for emergence of weeds. In relation to light regime, weeds may be classified into sciophytes – plants developing in the shadow in weakly thinned forest stands or in dense forest stands and represent no threat to tree development; semisciophytes – semi-shadow plants that develop in thinned stands and can do a lot of harm; and heliophytes – plants of open habitats such as clearings, strips, burnt areas, etc., and represent a big threat to renovation and development of trees. There are a number of other weed classifications due to their adaptation to abiotic factors such as water regime, temperature, physico-chemical soil characteristics, etc. during their evolutionary development. However, very important weed classifications in forestry, which would have practical significance from the aspect of weed control, are the following weeds of forest nurseries and weeds of forest plantations and forest stands.

The most Important Weeds in Forestry

Weeds in Forest Nurseries

Weed flora in forest nurseries differ from those found in forest plantations and forest stands. Given the extent of care measures applied, weeds in forest nurseries are very similar to those found in cultivated crops. They are mostly annual and perennial herbaceous weedy species. The most common grass weed species present in the forest nurseries include: Sorghum halepense, Cynodon dactylon, Alopecurus myosuroides, Digitaria sanquinalis, Echinochloa crus-galli, Poa annua, and Setaria spp. Dominant broadleaf species include: Amaranthus retroflexus, Ambrosia artemisiifolia, Chenopodium album, Cirsium arvense, Convolvulus arvensis, Erigeron canadensis, Datura stramonium, Galium aparine, Solanum nigrum, Sinapis arvensis, and Poligonum spp.

Control of weediness in forest nurseries is very important and quality planting material is the basic prerequisite for success in forest stand establishment. Since weeds are one of the most limiting factors for the success of nursery production, their control should be approached very seriously.

Weeds in Forest Plantations and Forest Stands

Weeds in forest plantations and forest stands differ from those in forest nurseries, because, in addition to different care measures applied in plantations and stands, the conditions in habitats also differ. Apart from ferns, herbaceous annual and perennial weeds, woody weeds such as shrubs, bushes, and shoots from the stumps of different tree types may also be present in forest plantations and stands. Woody weeds are very hardy and have a great power of regeneration; it is practically impossible to destroy them completely by mechanical means. The most common weed species present in forest plantations and stands are: Ambrosia artemisiifolia, Amorpha fruticosa, Asclepias syriaca, Erigeron canadensis, Solidago gigantea, Sorghum halepense, Sambucus nigra, Stenactis annua, Pteridium aquilinum, Rubus caesius and etc.

Weed Control in Forestry

There are numerous measures and procedures for weed control in forestry today, but, in order to fight weeds successfully, they should consist of different care and control measures. Described below are the six classifications of weed control measures.

Preventive Measures

The main goal of preventive measures is to prevent weed distribution. All measures used to protect any surface from weeds, i.e. to prevent weed seed growth in the field are considered preventive measures. Preventive measures in forestry weed control include:

- Control aided by sowing only pure crop seeds, which prevents spreading of weeds over sown surfaces.

- Destruction of weeds on non-agricultural areas; weeds that present a constant source of weediness and transportation of seeds to arable lands are developed on such areas.

- Prevention of the spread of weed seed by human activities by keeping agricultural and forest machinery and objects clean.

- Allelopathy is a phenomenon, where cultivated plants secrete exudates affecting the suppression of weeds. It is manifested in such a way that in the presence of certain plant species, many others are not able to thrive, or are slowly developed.

Mechanical Measures

Mechanical measures for combating weeds include basic treatment such as ploughing, disking, tilling and etc. Also regular measures in forest nurseries and plantations are hoeing and farrowing, undertaken during the greatest part of the vegetation period and especially emphasised during the entire spring and in early summer.

One of the ways of suppressing the already growing weeds and preventing their seed dispersal is mowing. Multiple repetitions exhaust the stored substances in the root and the plant is killed. In addition to mowing, one of the methods of weed suppression in forestry is also the pruning of shoots and stump shoots. However, this weed suppression method is relatively expensive due to intense labor and if repeated pruning is required depending on the weed species present. Concerns about increasing pesticide use have been major factors for research in physical weed control methods in Europe.

Physical Measures

Physical weed control measures applied in forestry involve the use of flame and superheated steam. Destruction of weeds by flame can be applied in forest plantations with wider spaces between the rows, provided that the crops are previously protected by metal shields. Burning weeds is carried out on non-productive areas such as forest railways, roads, and canals. Destruction of weeds using steam is applied in forest nurseries in preparation of substrates used for sowing or planting. This is also a form of sterilization which destroys weed seeds in addition to plant diseases and noxious insects. Orloff and Cudney believe that the use of flame for the reduction of weeds is the best at the end of growing seasons, because in this way destroy most weed seeds that are dispersed on the soil surface.

Mulches

The covering of soil with a variety of materials such as straw, stubble, polyethyelene films, and others, to prevent the emergence of weeds is utilzed on smaller areas, mostly in forest nurseries. Polyethylene films of varying colors and thickness are most often used. This type of weed control is efficient for annual weeds but has no effect on control of many perennial weeds, and can be expensive compared to other methods used to fight weeds.

Many types of mulches have been tried including: sheets of plastic, newspaper, plywood, various thicknesses of bark, sawdust, sand, straw, sprayed-on petroleum resin, and even large plastic buckets. Most have proven to be ineffective, costly or both. Early trials tended to use small, short-lived materials that aided conifer seedling survival but not growth. Compared to other weed control techniques available in previous years, mulches were rather expensive. Current trends are to apply longer-lived, somewhat larger mulches of mostly sheet materials made of reinforced paper, polyester, or polypropylene.

Biological Weed Control

Biological measures of weed control are based on the application of natural weed enemies such as insects, fungi, viruses, and bacteria in order to prevent their dissemination, and thus spreading. There are numerous examples of successful biological weed control. Application of pathogenic fungus, Chondrostereum purpureum, is used to control beech, yellow birch, red maple, sugar maple, trembling aspen, paper birch, and pin cherry. Exotic leaf pathogens, Phaeoramularia sp. and Entyloma ageratinae, were used for control of Ageratina adenophora and Ageratina riparia in South Africa. Gordon & Kluge mentioned that control of Hypericum perforatum can be done by using insects Chrysolina quadrigemina and Zeuxidiplosis giardi. For control of Acacia longifolia, the widely spread invasive plant species in Portugal, the bee wasp Trichilogaster acaciaelongifoliae was used. In those parts of the world where Eucalyptus sp. presents a problem the pathogen, Cryphonectria eucalypti, may be used for its suppression.

Application of biological measures in weed suppression has its limitations, though it has several advantages. Cultivated plants can be protected from some weeds, but not from all of them. It is impossible to destroy weeds completely because the biological agent depends upon the weed for survival; moreover, it is difficult to program biological protection for numerous cultivated plants from weeds with certainty since there are many similarities between weed species and cultivated plants.

Herbicides

Herbicides are used in forestry to manage tree-species composition, reduce competition from shrubs and herbaceous vegetation, manipulate wildlife habitat, and control invasive exotics. Unlike agriculture, the use of herbicides in forestry began much later and generally the application of herbicides in forestry was based on experiences from intensive agricultural production. The results of research in agriculture are applied in forestry with major or minor delays. Due to the lack of labour, high labour costs, and large areas, producers are more often interestedin the use of herbicides. Use of herbicides in forestry decreases weediness, particularly at the initial stages of development of forest nursery plants, when the effect of weeds on plants is the greatest; at the same time, much better economic efficiency in the production process is achieved. Also, possible mechanical damages to the nursery plants can be avoided, and it happens very often that any kind of mechanical treatment is prevented in early stages of plant development due to high soil humidity. Use of herbicides to control competing vegetation in young forests can increase wood volume yields by 50– 150%.

Beneficial Weeds

It is a well-known organic gardening fact that there are numerous plants and herbs that serve well as companion plants in order to keep pests at bay. These beneficial companions are typically planted on purpose in order to make a concerted effort at organic pest control. Seeds for these beneficial companions are found at most garden centers, and they are often considered to be "domesticated" varieties.

A beneficial weed, on the other hand, is a plant that is not typically thought of as being "domesticated," but it does still serve as a beneficial companion to your vegetables and flowering plants. Since a large number of gardeners are more concerned about pristine flower beds and lush green lawns that are free from the occasional pretty yellow dandelion, the majority of beneficial weeds are pulled up or poisoned with the harsh chemical weed killers.

The reality is that if these beneficial weeds were left in place, they would actually stave off hordes of unwanted pests and also help to keep the soil filled with valuable nutrients. With knowledge of what weeds might actually serve a positive purpose on your property, you can take your organic pest control and soil improvement efforts to extra lengths, all by allowing nature to work her magic.

Beneficial Weed Categories

Beneficial weeds typically fall into several categories, which can help you to define their best purpose on your own property.

- Repelling pests.

- Distractions or traps for pests.

- Soil fertilization.

- Ground cover.

- Natural herbicide.

- Attracting beneficial bugs.

- Useful for human use.

- Flavor enhancements.

Some beneficial weeds will actually serve multiple purposes, if you just know what to use them for. Once you know just how some of these beneficial weeds can improve your soil and your crop yields, you'll be a lot more likely to hesitate when you see a patch of dandelions growing in the middle of your tomato patch.

Pest Repellant

As organic gardeners, we know that planting onions and garlic near your lettuces, cabbages, carrots, and beets can help to keep pests away from your crops. The strong scents that onions and garlic plants put off will help to mask the delicious scent of your other crops, and thus protect them from being devoured.

Beneficial weeds work in much the same way—by masking the scent of the companion plants around them. Ground ivy, wormwood, wild oregano, and wild onions can work well to repel pests that would otherwise turn your tasty vegetables into their own dinner.

Some weeds are also useful to keep around because they have thorns or spines on them that can deter small animals. Rabbits and squirrels may seem impossible to deter at times, but you'd be surprised at just how effective a patch of thorn-covered weeds can be at keeping the critters away from your lettuce crop.

Pest Distractions

Insects track down their food by scent and color, which means that if you can effectively mask the scent of your tasty crops then you have won half of the battle. Some beneficial weeds work well as decoy plants that will not only attract the pests to them versus the tasty crops, but will also then often become the preferred meal over your crops.

Having a number of beneficial weeds like clover and ground ivy surrounding your crops has been shown to dramatically reduce numbers of tomato hornworms, squash bugs, and Japanese beetles. Patches of clover have also been known to distract the hordes of wild rabbits that would otherwise make short work of your cabbages and tasty young greens. Many an organic gardener has planted strips of clover as a type of organic line of defense. Keep in mind that some rabbits are just bound and determined to cross that line and make their way over to mow down the tops of your basil, but the clover should at least deter a fair number of them.

Soil Fertilization

We all know that planting beans amongst some of our other crops, like corn and squash, can increase the nitrogen content of the soil. Wild legumes can have the same effect on your crops, which means that white clover is actually a wonderfully beneficial weed to have growing in your garden.

Other weeds, like dandelions for example, have incredible root systems. They have extra-long tap roots that delve deep into the soil to bring up key nutrients that the other plants would otherwise not have access to. The quality of the soil will gradually improve over the course of a few years, but it would be constantly improving due to the presence of these soil-mining weeds. This also means that even if you have horribly dense clay soil, beneficial weeds like dandelions will ultimately transform the soil into something a lot more hospitable to vegetables and herbs.

Ground Cover

Think about the floor of a forest. This is one area that you would think would be densely populated with weeds and other unwanted plants. In reality, most forest floors are covered with thick ground-covering plants like ivy. Many beneficial weeds can be grown amongst your crops because they not only have different nutritional requirements from the soil, but they can also provide an effective ground cover that will essentially work as a type of living mulch. They will inhibit the growth of harmful weeds and will help to retain moisture in the soil. This can also often drop the temperature of the soil and help plants to better cope through the hottest and driest parts of the summer season. White and red clovers, along with rye grass, are often used to provide a nice ground cover for many crops.

Natural Herbicide

The plant world can actually be rather brutal, and the process of allelopathy is just one such example. Allelopathy is a process by which a plant produces biochemicals that will have a marked effect on the overall health of other plants around it. Allelopathy is not necessarily a positive process in all situations. For example, nut's edge can drive gardeners absolutely insane once it has been allowed to gain a foothold in the garden.

Not only it is difficult to control, but it also releases toxins into the soil, thus making the soil inhospitable for other plants. This type of process can be beneficial for other plants, however. For example, many consider lantana a weed, but it can inhibit the growth of milkweed roots.

Attracting Beneficial Bugs

A garden that is teeming with beneficial insects is a healthy garden! There are dozens of beneficial weeds that can attract beneficial insects to them by offering nectar and a much more hospitable place for them to lay their eggs. An influx of more beneficial insects can only be a positive thing, as they take over pest control efforts on your crops. Attracting more ladybugs and predatory wasps is always a good thing for an organic gardener.

Useful for Human Use

There are a number of uses for beneficial weeds, uses that go beyond the garden. Stinging nettles can certainly be irritating to your skin when you touch them, but you'll find that cattle and other livestock love eating them! As a matter of fact, you can eat them too! Nettles are incredibly high in a number of vitamins and minerals, including iron and calcium. They have a flavor that is very similar to spinach and can be cooked in much the same way. Nettles are often added to other herbs to create a vitamin-rich tea, so get a bit creative with your tea blends.

Dandelions are also definitely edible and are enjoyed in a number of dishes around the world. Dandelion leaves can be blanched to remove the bitterness, or they can be eaten fresh in salads and vegetable wraps. The petals of the flowers are used in the production of dandelion wine, while the roasted roots are often used to make dandelion coffee.

Purslane can also be enjoyed as a leaf vegetable, either in salads or stir-fried. All parts of the plant are edible, and purslane is often cooked much in the same way that spinach is cooked. It can also be added to soups and stews for the incredible nutritional content. Purslane is incredibly rich in omega-3 fatty acids and also contains high levels of iron, calcium, and vitamins A, C, and B.

Flavor Enhancement

Not only are some beneficial weeds edible in their own right, but studies have shown that a nice crop of stinging nettles planted amongst your mints and other herbs can actually increase the essential oil content of the herbs.

Dandelions have also been shown to improve the flavors of lettuces, tomatoes, and so many other vegetables around them. This can often be attributed to the Dandelion's ability to improve the nutrient content of the soil, however.

Aquatic Weeds

Aquatic weeds are unwanted and undesirable vegetation that reproduce and grow in water. If left unchecked may choke the water body posing a serious manace to pisciculture.

- They provide breeding grounds and harbour predatory insects.

- Provide shelter to predatory and weed fishes and molluscs.

- They restrict free movement of fry.

- They cause obstruction during netting.

- Limit living space for fish.

- Limit plankton production.

- Reduce sunlight penetration and nutrients.

- Upsets the equilibrium of physico-chemical properties of water.

- Cause imbalance in dissolved oxygen budget.

- Promote accumulation of deposits leading to siltation.

- Reduce water movement, thereby limits oxygen circulation in water.

- Some weeds release toxic gases that cause fish death and add foul smell to water.

Classification of Weeds

- Floating weeds: Eichornia, Pistia, Azolla, Lenpa, etc.

- Marginal weeds: Colecasia, Typha, Cyperus, Marsilia, etc.

- Emergent weeds: Nymphae, Myriophyllum, Nelumbo, etc.

- Submerged weeds: Hydrilla, Valisnaria, Chara,, Ceretophyllum, etc.

- Algal weeds: Spirogyra, Microcystis, Oscillatoria, Dinoflagellates, etc.

Specific Requirements

Specific requirements for commercial applicators in the aquatic pest control category relate generally to a practical knowledge in three areas. These areas may be defined as water use situations, the secondary effects of pesticides upon organisms in the aquatic environment, and the principles of limited area application.

Water use Situations

Habitats for aquatic weeds involve various proportions of water and soil, ranging from intermittently wet ditches to ditches which always hold standing water, to streams, stock ponds, farm ponds, lakes, and to intermediate habitats.

Static Water

Static water can be defined as water confined for considerable periods of the year, or totally confined within a known area with no movement of water to downstream locations. If a herbicide is applied for weed control, there is no reason to expect that any appreciable downstream effect may occur, except overflow resulting from unusual storm conditions. Water impoundments such as stock ponds, and in some cases farm ponds, will fit into this category.

Limited Flow Water Impoundments

This type refers primarily to farm ponds, lakes, and ditches. Ditches may be intermittently wet and dry, depending upon local climatic conditions. However, herbicides applied to habitats such as ditches may present some hazard to downstream locations, due to movement of the applied pesticide following an influx of water from surrounding areas. The purpose of the ditch is to drain the surrounding land area so considerable amounts of water must pass through the ditch area. In addition, many farm ponds may be characterized as having limited flow since there nearly always is an overflow pipe and an emergency overflow channel (spillway). The overflow pipe is designed to permit passage of a continuous and relatively well-defined amount of water at all periods of the year. The emergency spillway is provided to permit outflow of water from the pond at periods of the year when storm incidence may cause excess amounts of water to accumulate in the pond. In such cases, pesticide applications to limited-flow water areas may be found in small amounts in waters downstream from the application site. It is conceivable that larger amounts of pesticides from a treated area may be found downstream in the event of sudden rain storms, which interrupt or come immediately after pesticide application.

Moving Water

Moving water is characterized as water found in small streams, creeks, streams, and rivers where there is always some detectable movement. Applied pesticides may be found in downstream locations in varying~ amounts away from the area of original application. Such situations present the greatest potential for concern as an environmental hazard.

Secondary Effects of Pesticide Applications

Improper Application Rates

Proper application of herbicides to aquatic situations involves equipment calibration and calculation of appropriate water volumes in order to determine correct dosage rates. There are several well known and proven methods of equipment calibration and water volume calculation to determine pesticide application rates. Environmental hazard can result from the improper application rate.

Static Water

If application rates are too low in a static water situation, desired kill of pests may not be accomplished. However, the water supply may be contaminated and unsuitable for use by livestock or as an irrigation supply. In the event of excessive application rates, damage to the fish populations may result, either from direct toxicity or an excessively rapid kill of plant materials which may result in oxygen depletion in the water, leading to suffocation of the fish population. Excessive application rates might also exclude livestock from use of the water for a period of time, and would rule out the use of water supplies for irrigation for an indefinite period of time. However, little effect would probably be observed as far as downstream hazards are concerned, since little or no outflow normally occurs.

Limited-flow Water

Improper application rates could result in contamination of downstream water used by municipalities or communities for domestic water supplies. The hazardous condition would exist whether limited-flow water sources were treated with an application rate too low to accomplish a desired kill of vegetation or if the rate were excessive. It is conceivable that excessive rates might result in a too rapid rate of kill of vegetation which could lead to oxygen depletion and subsequent suffocation of the fish population. This might further complicate contamination of downstream water supplies utilized as domestic waters, due to bacterial contamination resulting from decay and decomposition of killed fish.

Moving Water

Application of pesticides to moving waters may lead to at least temporary contamination of downstream water supplies which may be utilized for domestic consumption. In addition, the pesticide, though applied locally for pest control, is certain to move to other areas of the stream and affect various aquatic organisms.

Faulty Application

There are two major hazards involved in faulty application of pesticides: (1) possible contact of applied pesticide with non-target organisms with resultant damage; (2) failure to apply the pesticide to the target pest, resulting in no kill of the desired pest. For example, it would be hopeless to apply granular herbicides in fast moving water, whereas they might work quite well in static water impoundments and even in limited-flow water situations. All currently registered herbicides employed for aquatic weed control are rated as limited, or not at all toxic to fish, birds, insects, and other aquatic organisms so long as proper application rates and techniques are employed. Other pesticide labels should be carefully observed to ensure that the aquatic environment is not unduly contaminated as a result of pest control efforts.

Limited Area Application

Aquatic weeds may occur in the whole body of water as submersed weeds, or may appear to cover the whole surface of the water as floating weeds. Conversely, the same weeds or other pests may occur only in limited areas within a body of water, whether it is a static, limited-flow, or moving body of water. "Limited area application" implies the advantage of improved safety to aquatic species, specifically the fish population. If pesticides that are potentially toxic to the fish population are applied to a limited area, the fish population can move to untreated water areas and escape potential toxic effects. Also implied in this concept is that a minimal amount of pesticide is applied, which tends to reduce the potential effect upon downstream environments in the event of spillover from the treated body of water.

Surface-applied Treatments

Contact pesticides are generally applied to control floating weeds. Usually only one-fourth to one-third of the surface area of the body of water is treated at a time to reduce the possible hazard of oxygen depletion resulting from too rapid kill of large masses of vegetation in the water, which may affect the fish population.

To determine the number of gallons of water in a pond or small lake, the following method may be used. First, determine the surface area. If the pond is circular, measure the radius in feet, square that figure and multiply by 3.1416; if the pond is rectangular, multiply the length by the breadth, in feet. Multiply the surface area by the average depth of water and finally, multiply this figure by 7-1/2.

Total Water Column Treatments

In this application technique, frequently employed with emersed weeds and often employed with algae treatments, the whole body of water (including the water column from the bottom of the water impoundment to the surface) is treated. The entire volume of the body of water is calculated and the chemical is added to reach a specified dilution in the total water column. An alternative is to calculate the entire water body and then treat only one-fourth or one-third of the total water column, based on surface area, confining the treatment to selected sections of the pond where the pest infestation may be more intense. Specific application techniques include injection directly into the water of the undiluted chemical, or arranging for some dilution of the chemical to be sprayed or cast upon the surface of the water. With either method, further dispersal throughout the water column is dependent upon water currents. Aquatic granules are formulated to provide rapid sink to soil-water interfaces to control emersed and submersed weeds.

Bottom Acre-foot Treatments

This is a specialized application technique which is intended primarily for control of submersed aquatic vegetation. A boat carrying application equipment drags a hose or boom over and just above the lake or pond bottom. The chemical is dispersed through nozzles and the specific gravity of the chemical causes the treatment to remain near the bottom and in the proximity of the rooted, submersed weeds. Fish can move out of this water level and avoid any direct contact with the chemical until chemical residues are diluted or dissipated.

Removal of Aquatic Weeds

All these weeds have to be eradicated using one or more of the following methods.

Manual Method

Manual removal of weeds involves physical removal of the weeds by hand. This may be practical if the pond is small and labour is cheap.

Mechanical Method

Some machines or implements are used for removing aquatic weeds. This method is normally applicable for larger water bodies. It is capital intensive and beyond the means of average fish farmer.

Chemical Method

- Weeds are eradicated using chemicals.

- Different weedicide are used for removal of different weeds present in aquaculture pond.

Common Weeds and the Weedicides for their Control

S.no.	Type of weeds	Herbicide	Dosage	Method of application
1	Water hyacinth	2,4-D	8 - 10 kg/ha	Foliar spraying
2	Ipomoea spp.	2,4-D	2 - 4 kg/ha	Foliar spraying
3	Sedges and rushes	2,4-D	5 -10 kg/ha	Foliar spraying/ root zone treatment
4	Lotuses and lilies	2,4-D	5 - 10 kg/ha	Root zone treatment
5	Ottelia, Vallisneria	2,4-D	10 -20 kg/ha	Root zone treatment

6	Aquatic grasses (in young stages)	Dalaphon	5-10 kg/ha	Foliar spraying
7	Aquatic grasses	Paraquat	2 kg/ha	Foliar spraying
8	Aquatic grasses	Diuron	4 kg/ha	Root zone treatment
9	Microcystis, other planktonic and filamentous algae	Diuron	0.1-0.3 ppm	Root zone treatment. Dispersal in water column
10	All submerged weeds	Ammonia	10-15 ppm	Root zone treatment . dispersal in water column
11	Pistia	Ammonia	1% aqueous solution with 0.25% wetting agent	Foliar spraying
12	Pistia	Paraquat	0.2 kg/ha	Foliar spraying
13	Salvinia	Ammonia	2% aqueous solution with 0.25% wetting agent	Foliar spraying
14	Salvinia	Paraquat	0.4 kg/ha	Foliar spraying

Biological Method

- The method is more advantageous since the undesirable weeds are converted into fish flesh.

- It is cheap as no labour is involved and most suitable from the social and environmental point of view.

- The method employs certain organisms which feed on the weeds.

- The grass carp which can eat up much more aquatic vegetation is itself an excellent example of biological weed control.

- The common carp helps in uprooting of certain plants.

- Tawes, Puntius gonionotus is also a good feeder of aquatic weeds.

- The Yamuna turtle consumes water hyacinth in the ponds.

Common Weed Eating Fish and the Weeds of their Preference

Fishes	Names	Feed upon
Grass carp	Ctenopharyngodon idella	Submerged weeds e.g: Hydrilla, Najas , Ceratophyllum, Potamogeton, Otteliaand duck weeds
Common carp	Cyprinus carpio	Tender shoots
Gaurami	Osphronemus goramy	Tender shoots of submerged weeds and filamentous algae
Pearl spot	Etroplus suratensis	Filamentous algae
Silver carp	Hypophthalmichthys molitrix	Algal bloom

Parasitic Weeds

Parasitic weeds of the families Orobanchaceae (Aeginetia, Orobanche, broomrape) and Scrophulariaceae (Alectra, Striga, witchweed) are considered to be among the most serious agricultural pests of economic importance in many parts of the world. The genus Striga includes about 40 species, of which 11 species are parasites on agricultural crops. The genus Orobanche has more than 100 species but only seven are considered as economically significant.

Geographical Distribution and Main Host Plants

Parasitic weeds have evolved specificity to crops and plants in the natural vegetation. Striga hermonthica (Del.) Benth., *S.* asiatica (L.) Kuntze and *S.* gesnerioides (L.) Vatke, in the given order, are the most economically important species in the semi-arid to sub-humid tropics. The former two species are almost entirely specific to grasses (cereals) such as sorghum (Sorghum bicolor (L.) Moench), maize (Zea mays L.), pearl millet (Pennsisetum americanum L.), rice (Oryza sativa L.), sugar cane (Saccharum officinarum L.) and others, while the third one is parasitizing dicot hosts, mainly cowpea (Vigna unguiculata (L.) Walp.), tobacco (Nicotiana tabacum L.) and sweet potato (Ipomea batatas(L.) Lam. Africa was described as the place of origin of the agriculturally important Striga species, particularly the Sudano-Ethiopia region, where also sorghum was postulated to be originated. *S.* hermonthica is widespread in the semi-arid zones of northern tropical Africa and it is also found in the south-western part of the Arabian Peninsula. S. asiatica, on the other hand, has a wide distribution in the eastern to southern part of Africa, Asia, Australia and the United States. The third species, S. gesnerioides, occurs in Africa, the Arabian Peninsula, the Indian subcontinent, and has been introduced to the United States.

The species of the genus Alectra are found mainly in tropical Africa and subtropical southern Africa. A. sessiliflora, and A. fluminensis are also found in subtropical Asia and tropical and subtropical South America, respectively. A. vogelii Benth. is the most important species parasitizing mainly grain legumes in sub-Saharan Africa, which include cowpea, bambara groundnut (Vigna subterranea (L.) Verdc.), soybean (Glycine max (L.) Merr.), mung bean (Vigna radiata (L.) Wilczek), groundnut (Arachis hypogaea L.) and common bean (Phaseolus vulgaris L.).

The Mediterranean region is considered to be one of the centres of origin of Orobanche species. The species are distributed worldwide from temperate climates to the semi-arid tropics. The distribution of Orobanche crenata Forsk. is restricted to the Mediterranean regions, the Middle East and East Africa (Ethiopia), while other species have a wider spread. Today, the species O. crenata, O. ramosa L., O. aegyptiaca Pers., O. cernua Loefl., O cumana Wallr., O. minor Sm. and O. foetida Poir. are one of the major biotic limiting factors to the production of legumes such as faba bean (Vicia faba L.), chickpea (Cicer arietinum L.), lentil (Lens culinaris Medick.), and to crops of the family Solanaceae [tomato (Lycopersicon esculentum Mill.), potato (Solanum tuberosum L.), and tobacco (Nicotiana tabacum L.)] and Asteraceae, mainly sunflower (Helianthus annuus L.).

Life Cycle

The seeds of the root-parasitic weeds vary in their ability to germinate immediately after they have reached maturity. Seeds of Striga and Orobanche are dormant and require a period of after-ripening or so called post-harvest ripening period, whereas seeds of Alectra vogelli can germinate immediately after harvest when germination requirements are met. Seed germination occurs when ripened seeds are preconditioned by exposure to warm moist conditions for several days followed by exogenous chemical signals produced by host roots and some non-hosts (germination stimulant). Upon germination, a germ tube, which is in close proximity to the host roots, elongates towards the root of the host, develops an organ of attachment, the haustorium, which serves as a bridge between the parasite and its host, and deprives it of water, mineral nutrients and carbohydrates, causing drought stress and wilting of the host. Stunted shoot growth, leaf chlorosis and reduced photosynthesis are also phenomena that can be observed on susceptible host plants which contribute to reduction of grain yield. Most of the seeds in the soil will not be reached by the stimulant, but will remain viable for up to 15 years, forming a seed reservoir for the next cropping seasons. The penetration of haustorial cells into host tissue (xylem and phloem system) is carried out mechanically by pressure on the host endodermal cells and by hydrolytic enzymes. Conditioning, germination, parasitic contact (attachment) and penetration are mediated by elegant systems of chemical communication between host and parasite. After several weeks of underground development the parasite emerges above the soil surface and starts to flower and produce seeds after another short period of time. Seed production is prodigious, up to 100 000 seeds or even more can be produced by a single plant and lead to a re-infestation of the field. Thus, if host plants are frequently cultivated, the seed population in the soil increases tremendously and cropping of host plants becomes more and more uneconomical.

Agricultural Significance and Yield Losses

A considerable loss in growth and yield of many food and fodder crops is caused by root-parasitic flowering plants. Globally, Striga have a greater impact on human welfare than any other parasitic angiosperms because their hosts are subsistence crops in areas marginal for agriculture. In general, low soil fertility, nitrogen deficiency, well-drained soils, and water stress accentuate the severity of Striga damage to the hosts. These are typically the environmental conditions for Striga-hosts in the semi-arid to subhumid tropics. Nowadays, Striga is considered as the greatest single biotic constraint to food production in Africa, where the livelihood of 300 million people is adversely affected. In infested areas, yield losses associated with Striga damage are often significant, ranging from 40-100 percent. Moreover, it is predicted that grain production in Africa is potentially at even increasing risk in the future. This is because several factors that influence the occurrence and may accelerate the future spread and the infestation intensity of Striga species in agricultural cropping systems. These include the future adaptation of Striga to crops and to a wide ecological amplitude, and a drop in soil fertility in tropical soils. The significant yield reductions result in little or no food at all for millions of subsistence farmers and consequently aggravate hunger and poverty.

Alectra vogelii is a serious pest in cowpea production in Africa. The parasite infection did not decrease cowpea dry matter production, but it significantly altered dry matter partioning by increasing the proportion of root dry matter. Crop yield losses resulting from A. vogelii infestation range from 41 percent to total crop loss of highly susceptible cultivars. The yield reduction is mediated through the delayed onset of flowering, reduced number of flowers and pods, and reduced mass of pods and grain.

The damage caused by the parasites Orobanche on field and vegetable crops is significant in the Near East, South and East Europe and in various republics of the former Soviet Union. It causes yield losses ranging from 5-100 percent. For example, in Morocco, the infestation of O. crenata in food legumes caused yield losses of 32.7 percent on an average in five provinces in the year 1994, which was equal to a production loss of 14 389 tonnes (US$8.6 million. As a result of the complete devastation caused by Orobanche in many areas, production methods had to be modified and cultivation of some susceptible hosts had to be abandoned.

Control Methods: Possibilities and Constraints

Compared with non-parasitic weeds, the control of parasitic weeds has proved to be exceptionally difficult. The ability of the parasite to produce a tremendously high number of seeds, which can remain viable in the soil for more than ten years, and their intimate physiological interaction with their host plants, are the main difficulties that limit the development of successful control measures that can be accepted and used by subsistence farmers. However, several control methods have been tried for the control of parasitic weeds, including cultural and mechanical (crop rotation, trap and catch cropping, fallowing, hand-pulling, nitrogen fertilization, time and method of planting, intercropping and mixed cropping), physical (solarization), chemical (herbicides, artificial seed germination stimulants, e.g. ethylene, ethephon, strigol), use of resistant varieties, and biological. These methods of control were well reviewed by Parker and Riches, and recently summarized in Kroschel, and Omanya. At on-farm level, the management of parasitic weeds is still unsatisfactory since - with the exception of the use of glyphosate in faba bean to control O. crenata - present control methods are not efficient enough to control already the underground development stages of the parasites. At present, the restoration of infested fields can only succeed through the improvement of existing farming systems based on a sound analysis of the parasitic weed problem and the development of a sustainable long-term integrated control programme consisting of the more applicable control approaches that are compatible with existing farming systems and with farmer preference and income. The success of cultural measures becomes evident only in the long run and will not improve yields in the present crop, because of the long underground developmental phase as well as the high seed production and longevity. The income of the subsistence farmers is usually too low to justify the use of highly sophisticated technical inputs such as ethylene to trigger ineffective Striga seed germination, as used in North Carolina to eradicate S. asiatica, or with soil solarization. In addition to the cost, selectivity, low persistence and availability are major constraints that limit the successful usage of herbicides. In addition, the use of synthetic germination stimulants and application of high dosage of nitrogen fertilizer (more than 80 kg N ha^{-1}, mainly as ammonium sulfate or urea), are not readily applicable in African farming systems. Few resistant lines for some host-parasite associations were reported but resistance is often interfered by the large genetic diversity of the parasites. Recent successes have been achieved in biological control, but it has not led to practical field application owing to the difficulties associated with mass rearing, release, formulation and delivery systemss. Strategies will be proposed on how to utilize this progress to formulate successful control methods that are economically accessible and acceptable to subsistence farmers.

References

- Weed, plant: britannica.com, Retrieved 29 June, 2019
- What-is-a-weed, weeds, plant-problems: gardeningknowhow.com, Retrieved 30 July, 2019
- The-basics-of-beneficial-weeds, food: offthegridnews.com, Retrieved 31 August, 2019
- Aquatic-weeds-0, content: agropedia.iitk.ac.in, Retrieved 1 January, 2019
- CAT5, Hold, cat5, PAT, Ag: uky.edu, Retrieved 2 February, 2019

2
Types of Pests

Pest includes crops, livestock and forestry that impacts the human activities adversely. Ants, bees, fleas, myriapods, rodents, flies, bed bugs, wasps, etc. are some of its types. The topics elaborated in this chapter will help in gaining a better perspective about pests and their types.

Pest

Pests refers to harmful organisms that latch on to plants, rendering them unsuitable for harvest. While most of these organisms tend to be insects, some fungi or plants can also be classified as pests.

Every garden is prone to pests. Some organisms are harmless but the majority are detrimental to a plant's roots, leaves, and overall health. This is why it is important to carefully prune plants and adopt the necessary precautions for proper pest control.

If left untreated, pests can destroy an entire crop. It should be known that they also affect flowers, vegetables, and fruits. Some of these harmful organisms have also been known to spread to neighboring crops.

Pests can stunt a plant's growth, especially smaller ones, disturb the soil, harm the foliage's appearance, and drastically reduce the overall quality of fruits and vegetables. When left untreated, pests often result in the plant's death.

Ants, snails, slugs, caterpillars, spider mites, whiteflies, fungus gnats, and aphids are the most common types of garden pests.

As far as pest control is concerned, the majority of gardeners ultimately resort to the use of pesticides. In some cases, these can harm the actual plants and disrupt the overall garden ecology, which is why some people choose to use organic products.

Pests include animals which:

- Carry disease-causing micro-organisms and parasites, for example, mosquitoes which carry Ross River virus and Murray Valley encephalitis.

- Attack and eat vegetable and cereal crops, for example, caterpillars and grass-hoppers.

- Damage stored food. For example, rats and mice may eat grain in silos, rice or biscuits in shops and homes and contaminate this food with their faeces (drop-pings) and urine.

- Attack and eat farm and station animals. For example, feral dogs (dingoes) kill or maim many sheep and goats each year; foxes will kill poultry, lambs and many species of native wildlife; and feral cats also prey on native wild-life.

- Damage clothing. Silverfish, for example, eat holes in clothes.

- Damage buildings. For example, termites can cause considerable damage to timber in buildings.

- Bite people. For example, bed bugs (so called because they often bite people in their beds) are very difficult and expensive to control. Their bites can cause great irritation to those bitten and, like mosquito bites, can become infected if scratched.

There are thousands of different kinds of pests which are harmful to humans. The great majority of these are types of insect.

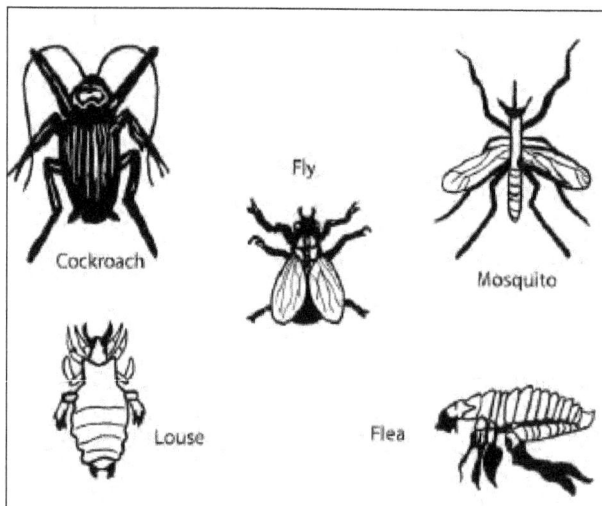

Agricultural Pest Insects

Pest insects can have adverse and damaging impacts on agricultural production and market access, the natural environment, and our lifestyle. Pest insects may cause problems by damaging crops and food production, parasitising livestock, or being a nuisance and health hazard to humans.

Grape Berry Moth

The grape berry moth is a key pest of grapes that is distributed in the United States east of the Rocky Mountains, and in eastern Canada.

The larvae of this insect can cause serious damage to commercial vineyards by feeding on the blossoms and berries. Infested berries may appear shriveled with fine webbing. Damage by grape berry moth may increase mold, rots and numbers of fruit flies. While grape berry moth larvae may only damage a few berries in a cluster, it is impractical for growers to remove damaged berries and webbing from clusters. Hosts include wild and cultivated grapes.

Grape Berry Moth Damage.

The adult moth is small, active, and about 1/4 inch long. When it is at rest with its wings folded, there is a brown band across the middle of the insect, the hind portion is gray-blue with brown markings, while the front portion is gray-blue without markings. The full grown larva is 2/5 inch long, pale olive-green, and can have a purplish tinge from the food it has eaten. The pupa is about 1/5 inch long, greenish-brown to dark brown and found under a flap cut in the leaf surface.

The grape berry moth overwinters as a pupa in leaf litter under vines. Adults begin to emerge in late May and lay eggs of the first generation singly on fruit stems just before blossom time. Eggs hatch in about 5 days. Under a flimsy web, the larvae feed for about 21 days on the blossoms and young fruit. In mid to late July, larvae move to leaves where they make a semi-circular slit, fold the flap over themselves and pupate. Adult moths emerge from the pupae in 10 to 15 days. Moths begin laying eggs for the next generation after 4 to 5 days. There may be 2 or 3 generations per year. Larvae of the second and third generations enter berries and feed within, passing from one berry to another under protection of webbing. Some of the cocoons of the second or third generations fall to the ground where they overwinter.

Webbing over blossoms and berries, and leaf flap cocoons are indicative of grape berry moth. In winter, the cocoons may be found in leaf litter under the vines. Clean up or bury leaf litter under vines in winter to eliminate over wintering pupae. Although larvae first appear when the grapes are in bloom, insecticides should not be applied until the berries are the size of small peas so as not to destroy beneficial pollinators. Insecticidal control of second generation is more difficult due to an extended flight period of moths as well as the difficulty of getting adequate spray coverage inside the cluster as berry size increases.

Pheromone traps are available to monitor for adult moth activity and enhance timing of insecticides for grape berry moth control. Recent studies in some states have shown mating disruption with synthetic pheromones to be an effective alternative in situations where there is no immigration of moths from outside sources. Mating disruption relies on releasing enough of the pheromone in the vineyard so that males cannot find female moths. Pheromone is imbedded in 8-inch plastic twist-ties using 400 twist ties per acre. Commercial systems available for mating disruption for this insect are recommended for vineyards at least 5 acres in size.

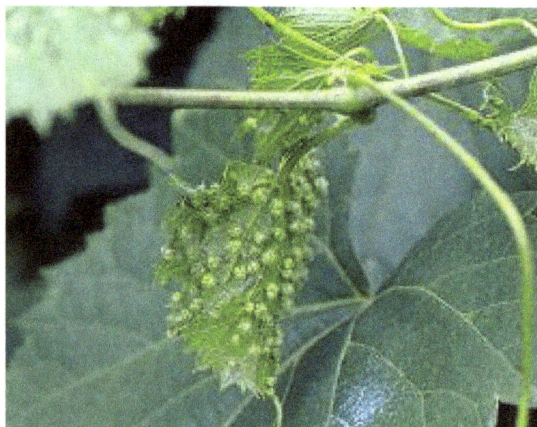

Grape Phylloxera.

Grape Phylloxera

Grape phylloxera is native to eastern United States, but has been distributed to other grape regions of the U.S. and is also established in Europe where it is of great economic importance. The leaf galls caused by grape phylloxera are unsightly and do little damage, however, infestation of the roots can be difficult to control and can lead to decline of vines. Severe infestations can cause defoliation and reduce shoot growth. Hosts include cultivated and wild grapes.

The wingless forms of the insect are very small, yellow-brown, oval or pear-shaped, and aphid-like. The winged forms, which are less apt to be seen, are also aphid-like, except that wings are held flat over the back. Neither winged nor wingless forms have cornicles, tail pipe-like structures on the top of the abdomen, as aphids do.

The presence of grape phylloxera is best recognized by characteristic galls it produces on the leaves or roots. Leaf galls are wart-like, about 1/4 inch in diameter, and are familiar to anyone growing grapes. Root galls are knot-like swellings on the rootlets, and may lead to decay of infested parts.

Root galls cause stunting and death of European varieties of grape vines. American varieties of grapes, or European grapes on American root stock are tolerant to the root gall form of the insect. Some varieties are resistant are to root galls, leaf galls or both.

The life cycle of grape phylloxera is complex due to the fact that generations with different life cycles may develop at the same time, at least in the eastern US. In spring, a female hatches from a fertilized egg that had been laid on the wood of a grape vine. She migrates to a leaf where she produces a gall and grows to maturity in about 15 days. She fills the gall with eggs and dies soon afterward. Nymphs that hatch from these eggs escape from the gall, and wander to new leaves where they in turn produce galls and eggs. There maybe 6 or 7 generations of this form during the summer. In the fall, nymphs migrate to the roots where they hibernate through the winter. The following spring they become active again and produce the root galls on susceptible varieties of grapes. These wingless females may cycle indefinitely on the roots year after year. In late summer and fall, in the eastern U.S., some of the root inhabiting phylloxera lay eggs that develop into winged females. These females migrate from the roots to the stems where they lay eggs of two sizes, the smaller ones developing into males and the larger ones into females. Mating occurs and the female then lays a single fertilized egg that over winters on the grape stem. It is this egg that gives rise to leaf inhabiting generations. Phylloxera cycle continuously as root inhabitants. Although they can cycle continuously on the roots without leaf forms occurring, leaf inhabiting forms do not occur without the root form also occurring.

European varieties of grapes should be grafted onto American grape root stocks. Foliar sprays to control phylloxera during their wandering stage do not accomplish any useful purpose.

Grape Rootworm

Grape rootworm is distributed from the Mississippi River eastward. Larvae devour small roots and pit the surface of larger roots, causing an unthrifty condition of the plant, and reduction in yield. Vines may be killed in 3 or more years when damage is severe. Adults make chain-like feeding marks on leaves and may also feed on the surface of green grape berries. Hosts include wild and cultivated grapes.

The adult beetle is elongate oval, sub-cylindrical, dark reddish brown, clothed with short pubescence and is about 1/4 to 1/3 inch long. The larva is white, hairy, curved, with a brown head. Grubs in various stages of development pass the winter in soil at a depth of a few inches to 2 or more feet. In the spring, they migrate to within 1 or 2 inches of the soil surface, where they root feed for a while before forming a small earthen cell to pupate in late May and June. Adults emerge over a period of 4 to 6 weeks, beginning about 2 weeks after grapes bloom, and feed on leaf upper surfaces in a characteristic manner. Soon after feeding begins, females lay eggs in masses of 20 to 30 on canes, usually under loose bark. Eggs hatch in 1 to 2 weeks, larvae drop to the ground, and enter the soil where they feed until the approach of cold weather. There is one generation per year.

Chain-like leaf feeding damage by the adults is diagnostic and can alert growers to adult activity. Foliar sprays when adults are active can provide effective control.

Grape Flea Beetle.

Grape Flea Beetle

Grape flea beetle is found in the eastern two-thirds of the United states. Adults eat buds and unfolding leaves, causing leaves to be ragged and tattered. Larvae feed on flower clusters and skeletonize leaves in a manner similar to adult rootworm feeding. Hosts include grape, plum, apple, quince, beech, elm and Virginia creeper.

Adults are dark metallic greenish-blue, jumping beetles about 1/5 inch long; larvae are brownish and marked with black spots, eggs are pale yellow and fairly conspicuous on upper leaf surface or under loose cane bark.

Grape Flea Beetle Larva.

Adults overwinter in protected areas around vineyards, and start feeding on interior of primary buds and opening grape leaves in early spring. Damaged buds will not develop into primary canes which can reduce yields. Once the buds are 1/2 inch long, only slight injury is caused. The females lay eggs under loose cane bark on vines, or occasionally on leaf upper surfaces, or on buds. The light yellow fairly conspicuous eggs hatch in a few days. The larvae feed on leaves for 3 to 4 weeks, then drop to the ground where they pupate in the soil, and emerge as adults in 1 to 2 weeks later. New adults feed for the remainder of the summer and go into hibernation in fall. There is only 1 generation per year.

Damage is often restricted to vineyard borders, particularly near wooded areas. Scheduled sprays for grape berry moth and leafhoppers provide effective control. Where flea beetles have been a problem, a spray timed at bud swell can provide control.

Grape Cane Girdler

Grape cane girdler is common in central and eastern United States. Adults girdle canes with a row of punctures, that causes canes to break off at the girdled areas. It is only a minor pest on grape, preferring Virginia creeper. Hosts include grape and Virginia creeper.

The adult is a black snout beetle about 1/8 inch long. The grub is slightly larger when full grown, and is white with a brown head and legless. It is very similar in appearance to the closely related grape cane gall maker.

Eggs are laid in late spring in a series of holes encircling the cane made by the female using its mouthparts. After eggs are laid, the female continues to make another series of punctures a few inches below the first girdle until the cane is encircled, but eggs are placed only in the holes of the first girdle. A similar girdle is made at a point higher on the cane, causing the end to break. Grubs feed in the cane pith between the girdles. After larval development is completed, pupation occurs. Adults appear in late summer, go into hibernation, and reappear in late spring.

Girdles are usually beyond the fruit clusters and do not cause significant yield loss. Look for broken off, pencil-sized canes with a grub in the pith of each broken off section, or wilted canes with a series of punctures. Pruning canes a few inches below the lower girdled area is usually sufficient control for this pest. Pruning should be done before adult emergence in late July or August.

Grape Cane Gallmaker

Grape cane gallmaker is a common pest of grapes in Kentucky. This insect produces noticeable red galls on new shoot growth just above nodes. While these are commonly found in vineyards, the majority of the galls are beyond the fruit clusters and usually cause no serious yield loss. Canes with galls are capable of producing a crop the following year.

The adult is a dark brown snout beetle about 1/8 inch long and is very similar in appearance to the grape cane girdler. The grub is slightly larger when full grown, and is white with a brown head and legless.

In May and June, the female lays an egg in one of a series of holes she chews along the cane just above a node when canes are 10 to 20 inches long. The larva feeds inside the cane which has developed a reddish swelling 3/4 to 1 inch long. The larva becomes fully developed in late July and pupates within the gall. Adult beetles emerge during August and remain in or near the vineyard area, overwintering in trashy borders.

Galls are usually found along vineyard borders near wooded trashy areas or at the ends of rows. If galls will be removed by pruning, it should be done by mid-July before emerging adults exit galls.

Grape Root Borer

Grape root borer is potentially the most destructive insect attacking grapes in Kentucky. Larvae of this insect tunnel into the larger roots and crown of vines below the soil surface. Borer damage results in reduced vine growth, smaller leaves, reduced berry size, and fewer bunches of grapes. Because damage is restricted to below ground, problems often go unnoticed until vine decline is observed. Damage caused by larval feeding can range from just a few feeding sites to complete root system destruction.

Adults are brown moths with thin yellow bands on the abdomen and resemble some paper wasps. The front wings are brown while hindwings are clear. Male moths fly about in a manner similar to wasps. Larvae are cylindrical, cream-colored, with three pairs of true legs near the head and five pairs of fleshy abdominal prolegs each bearing two bands of tiny hooks. The larvae are 1-1/2 inch long when mature and have a retractable brown head.

Adults emerge from the soil in mid summer. Eggs are laid on the soil surface, grape leaves, and weeds with eight days of adult emergence. Females lay an average of 350 eggs. Eggs hatch in about two weeks and larvae immediately tunnel into the soil in search of grape roots. About 95% of the larvae die before reaching roots, but less than 1% die after finding grape roots. Larvae will feed on the roots for 2 years. During the summer of the second year, larvae will pupate near the soil surface.

Injury by root borers is often most severe in low, poorly drained areas of the vineyard. In mid-summer, growers should examine around the bases of vines out to a distance of 18 inches for empty pupal skins of grape root borer. If pupal skins are found beneath 5% of the vines then an insecticide application is recommended next year underneath vines. It is best to apply the insecticide just as the adults are beginning to emerge, but the preharvest interval may make it necessary to spray after harvest. The insecticide should be applied as a course spray to a 15 square foot area surrounding the vine. Treat with an insecticide only if necessary. If grape root borer is not a problem, there is no reason to risk destroying the natural control processes and increasing production costs.

Redbanded Leafroller

Redbanded leafroller is an occasional pest of clusters and fruits, and its symptoms are very similar to grape berry moth. Larvae of this insect will feed on both foliage and clusters. Unlike grape berry moth larvae, redbanded leafroller larvae do not crawl into the berry but remain concealed in webbing on the cluster stem and feed on the stem as well as berries. While redbanded leafrollers are numerous in Kentucky, they are only an occasional pest of grapes.

The adult redbanded leafroller is a 1/2 inch long reddish-brown moth with small areas of silver, gold and orange. The moth is recognized by the red band extending across the front wings when at rest. The larva is a small, yellowish-green, unmarked caterpillar. The head capsule is the same color as the rest of the body.

The redbanded leafroller overwinters as pupae in leaf litter on the soil surface. Adults emerge in April and begin laying clusters of eggs on canes. Larvae of this early generation feed on unfolding leaves and are not usually of major concern. Larvae of the second and third generations feed during the summer and are of economic importance due to berry feeding. Second and third generations are larger than the first, and egg laying occurs over an extended period of time.

Insect Pests of Ornamental Plants

Ornamental plants enrich our lives every day and improve our environment. Flowers, shrubs, and trees beautify our yards and parks, while houseplants add a pleasant living touch to our indoor environment. Perhaps you are one of the many people who find satisfaction in planting and caring for ornamental plants. If so, sooner or later you will be confronted with insects which threaten to ruin your plants and undo your hard work. Learning to identify pest insects is the first step toward an effective pest management strategy. Insect Pests of Ornamental Plants are Given Below.

Aphids

Aphids occur on most plants, but the most commonly attacked plants include crape myrtles, roses, and silver maples.

Aphids are around one-eighth inch long. Many different species of aphids occur on ornamental plants. All are small, soft-bodied insects with piercing-sucking mouthparts. Depending on species, their color may vary from green, yellow, or black to pink or red. Some species, known as wooly aphids, are covered with white, cottony strands of wax. One of the more distinctive characteristics of aphids is the presence of two elongate, tailpipe-like structures known as cornicles that protrude from the end of the abdomens of most species.

Damage

Aphids cause damage by sucking plant sap and by transmitting plant diseases. Although individual aphids consume very little, aphids reproduce rapidly and can occur in extremely high numbers. Many species reproduce by parthenogenesis, which means that female aphids do not need to mate to reproduce. Many species can complete a generation in as few as 7 days. Feeding is often concentrated on young, expanding leaf and terminal tissue, and the physical damage caused by large numbers of piercing mouth parts can result in distorted plant growth. Aphids excrete large amounts of honeydew, or undigested plant sugars. This honeydew accumulates on leaves, causing them to be sticky. Honeydew also results in the growth of black sooty mold fungi. Although sooty mold does not damage the plant directly, heavy accumulations of sooty mold are unsightly and interfere with photosynthesis. Honeydew and sooty mold will also accumulate on vehicles and lawn furniture located under aphid-infested trees.

Common Species

The crape myrtle is the only host of the aphid bearing its name. The crape myrtle aphid occurs on the undersides of leaves and can build to extremely high numbers on susceptible varieties. The wooly alder aphid is a common species on silver maples. River birches are commonly infested with an aphid (no common name) that causes leaves to become distorted and reddened. The Asian wooly hackberry aphid is a relatively new, nonnative species that commonly occurs on sugarberry trees. These insects, which are covered with a white, cottony material, cause accumulations of sooty mold on infested trees. Oleander aphids commonly attack oleanders and butterfly weeds (Asclepias) planted in butterfly gardens. Yellow pecan aphids and black pecan aphids, two common pests of pecans, are discussed in a following section.

Management

Many insects prey on aphids, with lady beetles and lacewings being the most common. Parasitic wasps and fungal diseases also have a big impact on aphid populations. Natural biological control is the most important means of controlling aphids, and outbreaks are more likely when previous insecticide sprays have disrupted natural control. Certain varieties of crape myrtles, especially the "indica" varieties, are more likely to develop aphid problems than others. Consider susceptibility to aphids when selecting varieties of crape myrtles for planting. Heavy concentrations of aphids occurring on terminal growth of smaller plants often can be washed off with water sprayed from a garden hose.

Control

Acephate, azadirachtin, imidacloprid, malathion, insecticidal soap, pyrethrins + canola oil.

Soil-drench treatments with imidacloprid can provide effective, long-term control of aphids, but such treatments are slow acting and must be applied before heavy infestations develop.

Silverleaf whiteflies. Note the numerous eggs and flattened, scale-like immatures, aswell as the white-winged adults.

Whiteflies

Whiteflies are most commonly seen on gardenias, Ligustrum, Chinese privet, and hibiscus.

Whiteflies are about one-sixteenth inch long and are related to aphids. Adult whiteflies are small, moth-like insects covered with a white, waxy powder. There are several different species, but all carry their white, powdery wings folded tent-like over their bodies. They are most commonly found on the undersides of leaves, but clouds of adults will fly around infested plants when disturbed.

Immature whiteflies are immobile, scale-like insects that feed on the undersides of leaves. They are flattened and oval-shaped and, depending on the species, may have waxy filaments protruding from their bodies. However, these traits are difficult to see without a hand lens.

Damage

Like aphids, whiteflies suck plant sap through piercing- sucking mouthparts. They are also similar to aphids in their tendency to build to high populations and their ability to produce large amounts of honeydew, which eventually results in sooty mold.

Common Species

Citrus whiteflies are commonly seen on gardenias (cape jasmine). Bandedwinged whiteflies can be identified by the two gray bands across each wing and occur primarily on hibiscus and other malvaceous plants. The silverleaf whitefly is an important pest of many vegetable and nursery crops and occasionally occurs on landscape plants. This species can be especially difficult to control.

Management

Avoid unnecessary insecticide treatments that can disrupt natural biological control.

Control

Acetamiprid, Azadirachtin, Dinotefuran, Imidacloprid, Insecticidal soap, Neem oil, Horticultural oil.

Dinotefuran and imidacloprid are useful for controlling whiteflies when applied as a soil drench. Acetamiprid is one of the more effective foliar sprays for whiteflies. When attempting to control whiteflies with foliar sprays, apply at least two successive treatments 5 to 7 days apart.

Several other whitefly insecticides are labeled for application only by licensed commercial applicators. Because of the efficacy of these products and the difficultly of effectively controlling whiteflies, homeowners may wish to contract with commercial applicators for whitefly control in difficult situations.

Maderia mealybugs. Note the small, yellow crawlers and the cottony egg sacs.

Mealybugs

Mealybugs occur on gardenias and a few other landscape plants.

Mealybugs are one-eighth to one-third inch long. They are soft-bodied, wingless insects that are related to aphids and whiteflies. One of the key characteristics of mealybugs is that their bodies are covered with a whitish or yellowish powdery wax material. The body is somewhat flattened and oval-shaped, and, depending on the species, there may be elongate, waxy filaments extending from the margins of the body. In most species, eggs are deposited in cottony egg sacs attached to the plant. Mealybugs are slow-moving insects.

The larvae of some species of lady beetles superficially resemble mealybugs. These lady beetle larvae are often found in association with infestations of aphids or mealybugs and are often mistaken for mealybugs. But these are predators that actively eat aphids and mealybugs. They move faster than mealybugs and have a distinct head.

Damage

Mealybugs are more commonly seen on plants grown indoors and in greenhouses, but they also occur on certain landscape plants and outdoor potted plants. Like aphids and whiteflies, mealybugs have piercing-sucking mouthparts and produce honeydew that supports the growth of sooty mold. Mealybugs often concentrate their feeding on young tissue in the terminals of plants, and heavy infestations can distort leaves and stems. The accumulations of wax, shed skins, and cottony egg sacs, combined with the resulting honeydew and sooty mold, are unsightly.

Management

Naturally occurring predators and parasites play a key role in keeping mealybug populations in check. On heavily infested plants, you can temporarily reduce populations by washing with a forceful water spray.

Control

Acetamiprid, Acephate, Imidacloprid, Dinotefuran, Insecticidal soap, Neem oil, Horticultural oil, Pyrethrins + Canola oil.

Tulip tree scales are soft scales that occur on yellow poplar and magnolia trees and produce large amounts of honeydew.

Imidacloprid and dinotefuran control mealybugs when applied as a soil drench. Acetamiprid is one of the more effective foliar treatments. Multiple applications, applied at 7-day intervals, may be necessary when using foliar sprays.

Scale Insects

Scale insects occur on camellias, hollies, magnolias, euonymus, and many other plants.

Scale insects are one-sixteenth to one-half inch long. They are unusual in that they are immobile for most of their lives and do not resemble other insects. Scale insects spend most of their lives underneath a hardened or soft waxy covering, with their mouthparts imbedded in the host plants. Newly hatched scale insects are insect-like and are known as crawlers. Scale crawlers quickly imbed their mouthparts into the plant and form the scale-like covering for which this group of insects is named. Scale insects are related to aphids, whiteflies, and mealybugs. There are two major groups of scale insects—armored scales and soft scales—and there are many different species within each of these groups.

Soft scales secrete a waxy covering that is firmly attached to their bodies. Soft scales produce large amounts of honeydew, which, in turn, can support the growth of sooty mold. In most cases, mature soft scales are usually much larger than armored scales, ranging in size from ⅛ to ½ inch across.

Common species of soft scales include tulip tree scales, which are commonly seen on yellow poplar trees, as well as on many of the small, deciduous magnolias and occasionally on Southern magnolias; pine tortoise scales, which occur on pines; and Indian wax scales, which occur on Indian hawthorns and many other shrubs. Crape myrtle bark scales are serious new pests of crape myrtles.

Armored scales secrete a waxy covering that is not attached to their bodies (although this is difficult to see without a microscope). Armored scales do not produce honeydew and are generally smaller—one-eighth inch or less—and more flattened than soft scales.

Common armored scales include tea scales, important pests of camellias and hollies; euonymus scales, which occur on many species of euonymus; white peach scales/false oleander scales, which occur on the leaves of magnolias and many other plants; and San Jose scales.

Damage

Depending on the species, scale insects may feed through the bark on twigs and limbs or on leaves. Heavy infestations of scale insects can be very damaging to plants, causing unthrifty growth, honeydew (in the case of soft scales), distorted growth, and even death of branches, limbs, or entire plants. Infestations of armored scales are easy to overlook because their coverings blend in with the bark.

Management

Predators, such as lady beetles and small parasitic wasps, often keep scale populations in check. Outbreaks are more likely to occur when this naturally occurring biological control is disrupted by insecticide sprays; but outbreaks can also occur for other reasons, such as plant stress, absence of natural control agents, or individual plants that are inherently susceptible. Applications of pyrethroid insecticides are especially likely to flare scale outbreaks. Some homeowner mosquito treatments containing pyrethroids are labeled for application to shrubs to control resting adult mosquitoes, but gardeners should be aware that their use increases the potential for scale problems. Also, scale outbreaks tend to be more common in areas where there is an active mosquito-fogging program, particularly on plants that are nearest the application route.

In some cases, you can use pressurized water sprays, with a fine spray at approximately 30 PSI of pressure, to dislodge eggs and female scales from plants. Such treatments also help remove accumulations of sooty mold. This method is most effective against large soft scales, such as wax scales, infesting thick-leaved species such as hollies and Indian hawthorns. Apply these treatments before bud break, when only mature leaves are present. Tender leaves and buds can be injured with excessive water pressure. Hand removal can also be a helpful tool in reducing soft scale numbers on small or lightly infested plants. Some plant species or varieties are more prone to having scale problems than others, and one way to avoid problems with scale insects is not using plants especially prone to scale infestations. Heavy outbreaks of scales are often naturally controlled, but this takes time. Such a cycle of scale outbreak followed by naturally subsiding populations is often seen with tulip tree scales on yellow poplars. However, heavy infestations of scales can kill or severely injure ornamental plants before natural control occurs.

Control

Azadirachtin, Dinotefuran, Imidacloprid, Acephate, Malathion, Carbaryl, Neem oil, Horticultural oil.

Scale insects can be very difficult to control. In cases where control of scale insects becomes necessary, there are three basic control options: horticultural oils, systemic insecticides, and crawler sprays.

Horticultural oils control scale insects and their eggs by suffocating them. Hence, oil sprays control only scale insects/eggs they contact directly, and getting adequate spray coverage is important when using oil sprays. Horticultural oils can be applied during the late winter, as dormant or delayed-dormant type sprays, as well as during the growing season. If used improperly, horticultural oils can injure plants, so be sure to read carefully and follow label directions. Because horticultural oils can be effective against all life stages of scale insects, they can be very useful tools for scale control. Examples of currently available horticultural oils include Hi-Yield Dormant Spray, Bonide All Seasons Horticultural & Dormant Spray Oil, Monterey Saf-T-Side, Fertilome Scalecide, Fertilome Dormant Spray & Summer Oil Spray, and Ortho's Volck Oil.Because they constantly feed on plant sap, scale insects are susceptible to certain systemic insecticides. These are insecticides applied to the soil or injected into the trunk and taken up in the sap of the plant, where they are consumed by the scale insects.

Imidacloprid (Ferti-lome Tree and Shrub Systemic Insect Drench and Bonide Tree and Shrub Insect Control are two examples) is available for homeowner use as a soil drench to control infestations of soft scales. Note that this product is specifically labeled for use against soft scales (the group that produces honeydew) but is not labeled for control of armored scales. This product can be a useful tool to control soft scales, especially when used in combination with other methods. Use rate depends on either height of shrub or cumulative trunk circumference. Keep in mind, however, that this product is slow acting, and it may take a month or more to begin seeing results. Apply this treatment according to label directions when plants are actively growing.

Dinotefuran is a soil-applied systemic insecticide that is especially useful against armored scales (the group that does not produce honeydew). Currently, dinotefuran is not available in small-package homeowner formulations, but there are commercial formulations that are not classified as restricted-use products and do not require special licensing to purchase (Safari 20 SG and Zylam are examples). Packages of such products can be costly and may contain more product than you need. However, in situations where a large number of plants require treatment, this may be the most cost-effective way to control armored scale infestations. Before purchasing such products, read a specimen label carefully to be sure you understand how to apply the product and how to determine the amount to use per plant. Then, count and measure the plants that require treatment, and make careful calculations to determine the amount of product needed. Be sure to read and follow label directions carefully. Measure carefully when mixing drench solutions; commercial formulations are concentrated. Use rate is based on inches of circumference around the main stem for trees, or on height in feet for shrubs. Dinotefuran is slow acting but can provide many months of residual control. Be sure to observe pollinator-protection requirements. Do not apply dinotefuran as a foliar spray to plants that are in bloom.

Crawler sprays are contact insecticide sprays that provide effective control of scale crawlers. Crawlers are soft-bodied and generally easy to kill with good coverage of a labeled insecticide. But such treatments are effective only if you apply them when scales are in the crawler stage. This can be difficult to do because eggs of different species of scales hatch at different times of the year. Two methods you can use to determine when crawlers first appear are frequently examining infested twigs with a magnifying lens and using double-stick tape wrapped around infested twigs. When properly timed, two or three successive sprays at 7- to 10-day intervals will kill most of the crawlers, breaking the life cycle of the scale insects on the infested plant. Timing crawler sprays is generally more difficult for soft scales, which have fewer generations per year than armored scales. Insecticides recommended to control scale crawlers include malathion or acephate.

Use a Combination of Methods

Because scale insects can be difficult to control, it is often necessary to use a combination of methods. On severely infested plants, one of the first steps is to prune out any severely damaged and infested limbs. This helps reduce the number of scales present and increases your ability to get adequate spray coverage. Getting thorough coverage is very important when attempting to control scales with insecticidal oils or crawler sprays. One problem with assessing progress with scale control is that dead scales look much like live scales, and dead scales may remain attached to the plant for quite a long time. With soft scales, you can often gauge progress by the presence of honeydew. As long as soft scales continue to produce honeydew, they are still alive. Because ants, such as fire ants and carpenter ants, actively tend and protect soft scales in exchange for the honeydew they produce, controlling ants can help control scales. Likewise, controlling scales can help control certain ant species.

Use a Professional

Some scale insecticides are labeled for application only by licensed commercial applicators, and commercial applicators have the equipment necessary to effectively apply these products. Homeowners may wish to contract with commercial applicators for scale control in difficult situations.

Other Scale Control Options

Two other options for scale control must be mentioned. One is the "do nothing" approach. In some cases, a plant may experience a heavy outbreak of scale insects that is eventually brought under control by naturally occurring parasites and predators. This often occurs with tulip tree scales on yellow poplars. But it takes time for the predators and parasites to find the scale infestation and control it. This is more likely to occur if the homeowner does nothing than if harsh chemical sprays are used. With this approach, there is the risk that the situation could get much worse, resulting in serious plant injury, before this natural control occurs.

Another option for dealing with scale insects is to replace the plant. This is a drastic step you should take only after careful assessment and consideration. Occasionally, a plant becomes so severely infested and damaged by scales that this may be the best option. Obviously, when replacing such plants, you may want to choose a different species or variety of plant less prone to scale problems.

Adult thrips are less than one-sixteenth inch long. This photo shows adults of
three species: tobacco thrips, flower thrips, and western flower thrips.

Thrips

Thrips occur on many plants but are most important on roses, tropical hibiscus, and laurels.

Thrips are very small, elongate insects that are no more than one-sixteenth inch long
when fully mature. Most adults have fringed wings they carry folded lengthwise over
their bodies, but these are evident only when viewed through magnification. Immature
thrips are usually light yellow to lemon-colored and are spindle-shaped.

Damage

Thrips can cause damage by feeding on leaves as well as by feeding on flowers. Their injury reduces the aesthetic value of the blooms of roses and other similar plants. Thrips
also injure the foliage of certain plants, causing the leaves to have an unsightly, bleached
appearance. A thrips feeds by punching plant cells with its needle-like mandible and
sucking up the resulting plant juices. This results in silvery or bleached damaged areas.
Because feeding is often concentrated on young, actively growing tissue, flowers and
leaves are often crinkled or distorted as they continue to expand after being damaged.
Thrips also transmit certain plant viruses.

Common Species

Western flower thrips and flower thrips commonly damage the blooms of roses, especially light-colored varieties, and blooms of other plants. Greenhouse thrips often cause
serious damage to the foliage of Grecian laurel. Chilli thrips is a serious new, nonnative
pest of roses and many other ornamental plants.

Control

Acephate, Azadirachtin, Malathion, Imidacloprid, Bifenthrin, Cyfluthrin, Cyhalothrin,
Permethrin, Spinosad, Insecticidal soap.

Soil-drench treatments of imidacloprid will help control foliage-feeding thrips, but thrips feeding in flowers are more difficult to control. During certain times of the year, large numbers of thrips migrate from maturing weeds and other hosts. Weekly sprays may be necessary to minimize damage to intensively managed roses during such periods of heavy migration.

These blotch-type leaf mines were caused by larvae of the black locust leafminer. These tiny beetle larvae feed on the inner portions of the leaf, leaving the upper and lower epidermis intact.

Leafminers

Leafminers occur most commonly on hollies, boxwoods, and azaleas; occasionally occur on other plants.

Leafminers are one-fourth inch long or less. The term leafminer describes any insect that completes at least a portion of its life by living and feeding inside plant leaves. In most cases, the larvae feed on the leaf tissue between the upper and lower epidermis of the leaf. Depending on the feeding habits of the particular species, the mines may appear as irregularly shaped blotches or blisters or as winding tunnels. There are many different species of leafminers, representing several different groups of insects. Some of the most common leafminer species are flies, but some caterpillars live as leafminers for at least the early portion of their lives, and some species of sawflies and beetles are also leafminers.

Damage

Damage is caused by the larvae, which destroy leaf tissue by mining in the leaf, reducing leaf area and interfering with nutrient translocation. Extremely heavy infestations can result in enough loss of leaf area to adversely affect plant vigor and health. Fortunately, this is not common, and most leafminer infestations do not seriously affect plant health.

However, even light leafminer infestations can cause plants to be unsightly, and damage to broadleaved evergreens may persist. This aesthetic injury is the primary damage leafminers cause.

Common Species

Holly leafminers are the larval stage of a small fly. Actually, several different species of leafminers attack hollies. Most have only one generation per year and overwinter as larvae or pupae within mines in the leaves. Adults emerge and lay eggs when new leaf growth is forming in the spring. Heavy infestations can result in severe aesthetic injury and leaf drop.

Boxwood leafminers are larvae of a small, gnat-like fly. The yellow to orange larvae overwinter inside the leaf mine, pupate in the spring, and emerge as adults in mid-spring.

Azalea leafminers are the larvae of a small moth. Newly hatched caterpillars feed as leafminers inside the leaves, causing blister-like mines, but older caterpillars exit the mines and feed as "leafrollers" or "leaftiers." Injury is concentrated on young leaves at the ends of stems. There are several generations per year, and heavy infestations can cause plants to be unsightly.

Citrus leafminers are a relatively new pest of citrus trees, which are often grown as landscape trees in the southern portion of the state. Citrus leafminers are the larvae of small moths. They cause long, winding mines in the leaves of many types of citrus trees, causing plants to be unsightly.

Because this insect is newly introduced, it has few natural enemies, and infestations are often heavy. Because citrus trees are grown both as landscape plants and food crops, it is important to be sure any insecticides are specifically labeled for that use.

Locust leafminers are the larvae of small beetles. They cause irregular, blotch-shaped mines on black locust trees. In some portions of the country, this insect occurs in such high numbers that black locust trees may be severely defoliated by midsummer to early fall.

Management

Predators and parasites often keep leafminer populations in check, and outbreaks are more likely when biological control is disrupted by previous insecticide treatments. Hand-picking mined leaves can be an effective method of managing light infestations of leafminers. This is especially true for holly leaf miners, which overwinter inside the leaves. Picking and destroying infested leaves destroys the insects while also improving plant appearance. Light pruning can help control azalea leafminers, especially when combined with a foliar insecticide treatment. English boxwoods are less susceptible to boxwood leafminers than are other types of boxwoods.

Control

Acephate, Imidacloprid, Spinosad.

Because the larvae live in a protected location inside the leaves, leafminers can be difficult to control. Systemic insecticides such as acephate or imidacloprid are normally most effective. Soil-drench applications of imidacloprid control some species of leafmining flies. Contact insecticide sprays must be applied during the time when adults are active and laying eggs. The objective is to control adults and establish an insecticide residue that controls newly hatched larvae as they are boring into the leaf. Several applications may be required to adequately cover the period of adult activity. Treatments containing the active ingredient spinosad are especially useful against leafmining caterpillars, such as azalea leafminers and citrus leafminers, and some formulations of spinosad are specifically labeled for use on home-grown citrus.

Sharpshooters are large leafhoppers (about ½ inch) that feed on twigs of crape myrtles and other plants, excreting large volumes of undigested plant sap.

Leafhoppers

Leafhoppers are minor pests of most plants.

Leafhoppers are one-eighth to one-half inch long, active, long-bodied, somewhat wedge-shaped insects. They have piercing-sucking type mouthparts and readily run, hop, or fly when disturbed. There are many different species; most are green to yellow, but some species are brightly marked with yellow, red, or blue.

Common Species

Glassywinged sharpshooters are common on crape myrtles. This is an unusually large leafhopper (approximately one- half inch long) that forcibly excretes large amounts of liquid. Potato leafhoppers can cause leaf injury on red maples, while the redbud leafhopper is a common inhabitant of redbud trees.

Damage

Both adults and nymphs feed on the undersides of leaves and on tender stems, sucking the sap and causing leaves to become spotted or turn yellow or reddish and dry up. In other cases, leafhopper injury causes distortion of leaves and terminals. Leafhoppers can also transmit important plant diseases, such as aster yellows and Pierce's disease of grapes. The feeding of some species causes a toxic response in the plant, resulting in a burning effect that can cause leaf tips to wither and die. In most cases, leafhoppers are minor pests that seldom cause serious plant injury.

Control

Bifenthrin, Cyfluthrin, Cyhalothrin, Malathion, Carbaryl, Imidacloprid, Permethrin, Insecticidal soap, Pyrethrins + Canola oil.

It is rather uncommon for leafhopper populations to become heavy enough to require treatment. Apply foliar sprays as needed. Soil-drenches of imidacloprid provide some control but are rarely applied specifically to control leafhoppers.

Eastern lubber grasshoppers can damage many types of ornamental plants, including thick-leaved plants such as amaryllis and lilies.

Grasshoppers and Crickets

Grasshoppers and crickets are general feeders on the leaves of many plants.

Grasshoppers and crickets can be one-half to 3 inches long. There are many different species of grasshoppers and crickets. They vary in size and color, but most are easily recognized by their hind legs, which are well developed for jumping.

Damage

Adults, as well as the immature nymphs, cause damage by eating leaves. Normally, the amount of defoliation these insects cause is negligible, but outbreaks occasionally occur and cause excessive defoliation. Potential for significant damage is greatest on small plants. In cases where grasshoppers are migrating from nearby fields, roadsides, or natural areas and feeding on ornamental plants, repeated insecticide treatments may be required.

Control

Carbaryl, Bifenthrin, Cyfluthrin, Acephate, Malathion.

Apply foliar sprays as needed to prevent excessive injury.

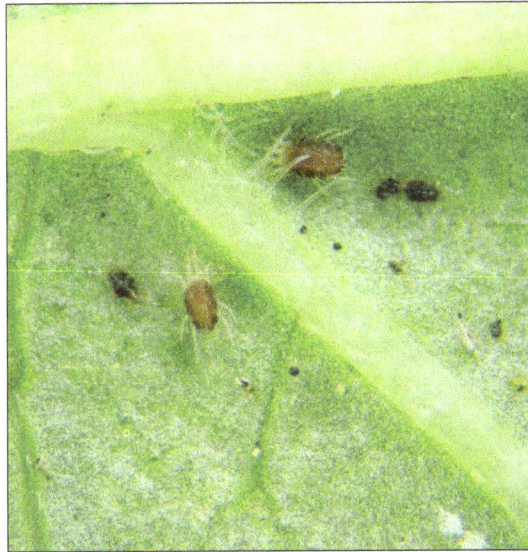

Spider mites are potentially damaging pests of most ornamental plants.

Spider Mites

Spider mites occur on roses, boxwoods, ornamental conifers, azaleas, camellias, and other plants.

Spider mites are one-thirty-second of an inch or less long. Although they are not insects, spider mites belong to a closely related group. Adult spider mites are so small that they are barely visible to the naked eye, but they can be readily observed through a 10x hand lens. Adults of most species are somewhat globular in shape and have eight legs. There are many different species, and color may vary from red to green or yellow. One of the more common species, the two-spotted spider mite, appears to have a dark spot on either side of its body.

Damage

Spider mites feed by sucking the fluid from plant cells. Adults and nymphs cause similar injury. Feeding by low numbers of mites is minor, but these pests have a very high reproductive potential and can complete a generation in as few as 7 days. Heavy infestations can cause severe injury and can even kill plants. Feeding by individual mites causes localized cell death, resulting in light-colored stippling.

When mite populations are heavy, these individual feeding sites fuse, giving leaves a bleached or bronzed appearance. Severely injured leaves may curl and drop from the plant. At first, mite infestations are just on the undersides of leaves, but, under heavy infestations, the mites produce webbing (hence, the name spider mite) and occur on the tops of leaves and on other plant parts.

Common Species

The two-spotted spider mite is a common species on many ornamental plants, including roses. Southern red mites commonly occur on hollies, camellias, and azaleas. Boxwood mites are common on boxwoods. Spruce spider mites are pests of many ornamental conifers.

Management

Naturally occurring predatory mites and other predators often keep populations of plant-feeding mites in check. Outbreaks of spider mites often occur following insecticide treatments targeted against other pests because these treatments destroy the predatory mites. Avoid unnecessary insecticide treatments. Foliar applications of carbaryl, acephate, or pyrethroid insecticides tend to trigger mite outbreaks, and outbreaks are also more likely to occur on plants treated with imidacloprid. Outbreaks of some species of mites are favored by hot, dry weather, especially if accompanied by dusty conditions. Thus, keeping plants well watered during periods of drought helps reduce the potential for mite outbreaks. Washing foliage with a water spray can also be beneficial in controlling or preventing mites, but this should be done early in the day to avoid prolonging leaf wetness, which encourages plant diseases.

Control

Malathion, Insecticidal soap, Neem oil, Horticultural oil, Pyrethrins + Canola oil.

Unfortunately, there are no specific, highly effective miticides that are sold as homeowner formulations. Horticultural oils can provide very effective mite control when thorough spray coverage is achieved. When treating for mites, it is important to apply two or more successive treatments 4 to 5 days apart to effectively break the life cycle. Choose the product carefully. Using products that don't work or inadequate treatment intervals can intensify mite problems. For serious mite problems on high-value plants, consider contacting a licensed commercial applicator. These professionals have access to more effective miticides.

Slugs can damage tender, low-growing plants such as hostas and begonias.

Slugs and Snails

Slugs and snails attack hostas and other tender perennials, as well as many annual plants.

Slugs and snails can be one-half to more than 3 inches long. Slugs are more closely related to oysters and other mollusks than to insects. Their long, fleshy, slime-covered bodies are not jointed, and they have two movable "eye stalks" and another pair of sensory tentacles on their heads. Although similar, snails have shells, whereas slugs do not. Both slugs and snails move about on a layer of slime that dries to leave shiny trails. There are several species of slugs, some of which may be longer than 3 inches when mature. There are many different species of snails, as well. Through most of Mississippi, slugs are generally more common and more damaging than snails. These animals are active mostly at night and spend the day hiding under flowerpots, mulch, leaf litter, or in other protected sites.

Damage

Although snails and slugs often feed on decaying plant material, they also damage the leaves of tender perennials and annual bedding plants by feeding on them with their rasping-type mouthparts. Damage appears as long, narrow holes in the leaves and is often associated with the shiny trails. These pests also damage blooms, especially blooms growing near the ground. Keep in mind that these pests are rarely active during the day, but their shiny foraging trails can indicate their presence. Feeding is often concentrated on the emerging leaves of plants like hostas, which magnifies the overall impact of the injury. High populations often occur in situations where there is heavy mulch, leaf litter or other organic matter and in densely planted beds. They thrive in conditions of constant moisture.

Management

Slugs thrive in moist, protected areas with heavy accumulations of decaying organic

matter. Limit conditions favorable to slugs, such as excessive moisture, excessive organic matter/mulch, excessive leaf litter and other detritus, and items such as flower
pots, rocks, and fallen limbs that provide daytime hiding places. Raking mulch away
from the base of susceptible plants can also help reduce attacks. You can use copper
or other types of barriers to protect especially sensitive or valuable plants. You can
control slugs through diligent use of traps baited with beer or other attractive baits,
such as moist dog food. Another nonchemical control option is to place inverted flower
saucers, boards, or other attractive harborages in the flower bed, check them regularly,
and physically remove and destroy any snails or slugs you find. These controls are more
practical for small plantings.

Control

Metaldehyde, Iron phosphate.

Slugs and snails can be controlled with baits specially formulated for this purpose, and
many commercial brands are available for home use. For best results, combine baits
with management efforts to make the area less attractive to slugs and snails. To prevent injury to developing leaves of hostas and other susceptible plants, it is important
to begin control efforts early in the spring before plants begin to break dormancy or in
the fall of the preceding year. Many species of slugs overwinter as adults and become
active as temperatures warm in the spring. Baits are generally less effective when slugs
are less active because of cool temperatures. Repeated bait applications are usually
required to obtain and maintain control. Place baits near areas where the pests hide.
Keep in mind that baits tend to lose their effectiveness when they become wet or if they
are exposed to sunlight for prolonged periods. Because baits containing metaldehyde
are toxic to pets and wildlife, be careful to use them properly. Baits containing iron
phosphate are labeled for use around domestic animals.

Pillbugs primarily feed on decaying plant material but sometimes damage
emerging leaves of hosta and other tender, low-growing plants.

Pillbugs

Pillbugs attack hostas and other tender perennials, as well as many annual plants.

Pillbugs are about one-fourth inch long. They are not insects but are land-dwelling crustaceans related to shrimp and crayfish. These animals have jointed bodies and seven pairs of legs and can roll into a ball when disturbed (hence, the name pillbug). Pillbugs prefer moist, protected environments with lots of decaying organic matter. Heavy populations often occur in heavily mulched flowerbeds or areas with heavy accumulations of leaf litter.

Damage

Although pillbugs feed mostly on decaying organic matter, they sometimes feed on the leaves of tender ornamental plants, such as hostas, and bedding plants. Damage often occurs as the leaves of dormant perennials are emerging through mulch and leaf litter.

Management

As with slugs, limiting mulch, leaf litter, and other detritus in flowerbeds can help limit pillbug numbers. Damage to emerging hostas and other tender perennials can also be limited by carefully raking mulch and leaf litter away from plants during emergence. This prevents the pillbugs from having a protected place to feed on the emerging foliage.

Control

Spinosad or Carbaryl baits, Bifenthrin, Permethrin, Cyfluthrin.

Pecan phylloxera are small, aphid-like insects that cause galls on leaves and twigs of pecans.

Baits containing spinosad or carbaryl help control pillbugs. These are usually sold as "bug and slug baits" and also contain active ingredients, such as metaldehyde or iron phosphate, that control slugs.

Liquid sprays or granular insecticides containing pyrethroid insecticides (active ingredients such as permethrin, bifenthrin, cyhalothrin, or cyfluthrin) can also be effective.

Pecan Phylloxera

Pecan phylloxera attack only pecans and hickories.

Although pecans produce edible nuts, they are also one of the more common trees in Mississippi landscapes. Pecan phylloxera is one of the most common insect pests of pecans. Because pecan trees are also food-bearing trees, it is important that you use only insecticides specifically labeled for application to pecans when it is necessary to treat them.

Pecan phylloxera are small, yellow insects that look very much like aphids. They are seldom seen because they are encased inside the galls that they cause to form on stems and nuts. The pea- to marble-sized, knot-like galls these insects cause make them easy to identify.

Damage

Pecan phylloxera overwinter as eggs in cracks and crevices on limbs and branches. Egg-hatching coincides with leaf bud break in the spring, and the young nymphs immediately crawl to the developing leaf buds and begin feeding. Their feeding affects the growth of the leaf tissue, causing the formation of the hollow, knotty galls that encase the feeding nymphs. These galls cause severe deformation of developing twigs and nuts. On heavily infested trees, more than 70 percent of the new terminal tissue can be affected, resulting in trees that are unsightly and unproductive. Fortunately, outbreaks of pecan phylloxera are somewhat cyclic, and trees may experience heavy infestations for a year or so, followed by several years of low populations.

Management

Avoid planting pecans in the home landscape. If you do plant pecans, avoid the high-maintenance varieties normally grown in commercial pecan orchards. Instead, plant varieties such as Candy, Elliott, Farley, Jenkins, or Syrup Mill that tend to perform better in unmanaged landscape situations. Although they generally produce smaller nuts, these varieties exhibit phylloxera- and disease-resistance or exhibit other traits desirable for unmanaged trees. Be aware that these varieties may be difficult to locate, but, if you wish to plant pecan trees in your home landscape, it is worth the effort to locate these or similar varieties.

Control

Carbaryl

You can control pecan phylloxera with well-timed foliar sprays of carbaryl, but few homeowners have the power- spray equipment needed to treat large pecan trees. In most cases, licensed commercial applicators must treat large trees, but not many are willing or able to do it.

Always be aware of the potential for drift onto neighboring properties and other non-target sites, and take appropriate precautions to avoid drift-related problems. In many urban settings, the potential for problems from spray drift may be so great that you won't want to make such treatments.

If you attempt treatment, proper timing is critical. Treatments will not be effective after the protective gall forms around the insect. To be effective, treatments must be applied as soon as leaf buds begin to break in the spring and before there is more than 1 inch of new leaf growth.

If the tree were severely infested the previous year, apply a second application about 10 days after the first.

Getting good spray coverage is also an important consideration when treating for pecan phylloxera. Depending on tree size, 10 to 20 gallons of finished spray are normally required to adequately treat one tree. When treating pecans, be sure the insecticide you use is specifically labeled for use on pecans.

Pecan Aphids

Pecan aphids attack only pecans.

Pecans are included in this publication because they are often grown in the home landscape. Because they produce an edible crop, be sure to use only insecticides specifically labeled for pecans.

Two species of aphids commonly occur on pecans. Their common names provide good general Descriptions.

Yellow aphids are small, yellow aphids that occur on the undersides of pecan leaflets. Yellow aphids can build to high populations, with numbers exceeding 50 to 100 aphids per compound leaf. Black aphids are small, black aphids that also occur on the undersides of leaflets, but they are much less numerous than yellow aphids. Black aphids are easily identified by the angular, yellow lesions that their feeding causes on pecan leaflets. Often, you can see an aphid feeding in such an area.

Damage

Yellow aphids cause damage by sucking plant sap and producing honeydew. The loss of sap and other associated damage can adversely affect vigor and nut production, and this is an important consideration for commercial producers. But for homeowners, honeydew and the resulting accumulations of sooty mold are often considered to be the more important injury. This is especially true for pecan trees growing over patios or parking areas. Although outbreaks can occur in late spring to early summer, heavy infestations of yellow aphids occur most commonly in late August through October.

Although they are much less abundant than yellow aphids, black aphids can cause more damage to the vigor and yield potential of pecans. This is because they inject toxic saliva that results in angular shaped, yellow lesions. These lesions eventually turn brown, and leaflets and entire compound leaves will be shed from the tree prematurely. Because they are much less numerous than yellow aphids, black aphids do not usually cause large amounts of honeydew.

Management

Avoid planting pecans in the home landscape, or at least avoid planting pecans near patio and parking areas where you don't want accumulations of honeydew and sooty mold on automobiles or lawn furniture. If you plant pecans, avoid planting high-maintenance varieties. Instead, plant varieties such as Candy, Elliott, Farley, Jenkins, or Syrup Mill that tend to perform better in unmanaged landscapes. Although they generally produce smaller nuts, these varieties resist or tolerate insects and disease or have other traits desirable for unmanaged trees. Be aware that these varieties may be difficult to find, but if you want to plant pecan trees in the home landscape, it is worth the effort to find them or similar varieties. Avoid spraying pecan trees with insecticides. Because many foliar insecticide sprays destroy predators and parasites that help keep aphid numbers in check, using them can cause aphid outbreaks.

Control

Imidacloprid

There are no insecticide sprays recommended to control aphids on pecan trees in the home landscape. Applying foliar insecticide sprays to pecan trees often increases aphid populations. The soil-applied systemic insecticide imidacloprid is labeled for use on homegrown pecans to control aphids. Because this product is applied as a drench to the soil around the tree and because it works systemically, it is less likely to disrupt natural control. Although control is often erratic, because of varying soil and weather conditions, it is the only useful treatment available to homeowners, and it is relatively easy to apply.

Azalea lace bug adults. Note the lace-like wings of the adults and the dark fecal deposits on the leaf.

Homeowners who choose to use the imidacloprid drench treatment should read the label carefully to be sure they are applying the proper rate. The product is sold in 1-quart bottles, and the use rate is based on the number of inches around the tree at breast height. It may take more than 1 quart of product to treat one tree. For example, a tree that has a diameter of 12 inches has a circumference of nearly 38 inches. It would take 38 fluid ounces of product to treat a tree of this size. Because it takes a long time for this systemic insecticide to be taken up and moved through the plant, you must apply it well before you expect a pest problem.

Lace Bugs

Lace bugs attack azaleas, lantanas, pyracantha, rhododendrons, sycamores, oaks, and a few other plants.

Adult lace bugs vary in color from brown to light gray, depending on species. They are about one-eighth inch long and have lace-like wings. In most species, the wings are enlarged and have net-like veins. Nymphs may be black to gray and are usually covered with spines.

Damage

Adults and nymphs cause damage by sucking sap from the undersides of leaves. Damaged leaves have a stippled appearance, which is often mistaken for spider mite injury. Heavily damaged leaves may look bleached out before eventually turning brown. Dark spots of shiny, shellac-like fecal material on the undersides of leaves are a sure sign of lace bug infestation and may be present even when insects are not obvious. Azalea, lantana, and pyracantha species are the most likely to require treatment. Although lace bugs are common on sycamores and some oaks, infestations are seldom severe enough to cause serious injury.

Common Species

Azalea lace bugs are the most common insect pest of azaleas. Heavy infestations can cause foliage to appear bleached out and unsightly. Because of the evergreen nature of azaleas, damage will remain for quite a while after insects are controlled. Lantana lace bug is a common pest of lantanas.

Heavy infestations cause bleaching and browning of leaf margins and eventually cause death of entire leaves. Lace bug injury on lantanas is often mistaken for drought stress, disease, spider mite feeding, or chemical injury. You must look closely to see the small insects on the undersides of the leaves.

Management

Lace bugs have several natural enemies. Check susceptible plants regularly to detect infestations before serious injury occurs. Azaleas planted in full sun are more suscep-tible to attack than those planted in filtered shade. Some varieties of azaleas exhibit resistance.

Control

Acephate, Malathion, Imidacloprid, Bifenthrin, Cyfluthrin, Cyhalothrin, Permethrin, Neem oil, Horticultural oil, Pyrethrins + Canola oil.

Soil-drench treatments of imidacloprid applied in fall or spring help control lace bugs. Foliar insecticide sprays are the quickest way to eliminate heavy infestations. Foliar insecticides that provide systemic activity, such as acephate or imidacloprid, are most effective. Direct insecticide sprays to the undersides of leaves for best control, especial-ly with nonsystemics.

Azalea caterpillars often assume this C-shaped posture when alarmed.

Azalea Caterpillars

Azalea caterpillars primarily attack azaleas, especially the "indica" varieties. They sometimes attack blueberries.

Mature caterpillars are approximately 2 inches long. This large, strikingly marked caterpillar is an occasional pest of azaleas throughout the South. Newly hatched azalea caterpillars are yellow with longitudinal reddish stripes, but their appearance changes markedly as they grow. Older caterpillars are black checkered with yellow or white, with reddish-orange heads and legs. These caterpillars often rest with their heads and tails raised into the air, creating a broad U-shape.

Damage

The moths deposit their eggs in masses of up to 100 eggs. Newly hatched larvae feed togeth-er on the undersides of leaves, causing leaf skeletonization. As larvae grow, they spread out and feed individually, causing progressively greater amounts of defoliation. Heavy infestations can cause total defoliation of entire plantings of azaleas. As with most caterpillars, 80 to 90 percent of the total leaf area that a single caterpillar will eat during its life is eaten during the last 3 to 4 days before pupation. This is why severe defoliation can seem so sudden. Plants that appear perfectly fine on Sunday afternoon can be totally defoliated by Wednesday afternoon. Infestations are most common on the large-leafed "indica" varieties.

Management

Be alert for early signs of defoliation injury to azaleas: leaf skeletonization caused by young caterpillars. Infestations are most common in late summer and early fall. Early detection and control of young larvae can prevent serious defoliation injury.

Control

Permethrin, Bifenthrin, Carbaryl, Acephate, Cyfluthrin, Cyhalothrin, Bts, Spinosad.

Azalea caterpillars can be controlled with foliar sprays containing these active ingredients. When treating for azalea caterpillars, you should also consider the potential for lace bug problems, and, if necessary, choose a treatment such as acephate or cyfluthrin + imidacloprid that controls both pests. Use Bt products only against small larvae not threatening to cause immediate defoliation.

Eastern tent caterpillars build angular-shaped webs in the crotches of black cherry tree limbs in the spring.

Eastern Tent Caterpillars

Eastern tent caterpillars occur primarily on black cherries but also attack apples, crabapples, and occasionally other trees.

Mature caterpillars are about 2 inches long. Eastern tent caterpillars are easy to identify by the silken tents they build in the crotches and limb forks of black cherry trees in early spring. The tents normally have an overall angular shape because of their location. Because few, if any, leaves are enclosed within the tent, the caterpillars must leave the tent, usually at night, to feed on nearby leaves. The background color of the body is black, etched with fine gold or yellow markings, but a distinct white line runs down the center of the back, and a row of irregular blue spots and markings run down either side. The head is black, and the body is sparsely covered with long, fine, tan-colored hairs. Mature larvae wander about on the ground in search of a place to pupate. Eggs are laid in dark-colored, somewhat slick, spindle-shaped masses around the smaller twigs. This is the overwintering stage.

Damage

Heavy infestations can cause significant, or even complete, defoliation, but there is only one generation per year, and trees can recover with little long-term injury. Protect young, newly established trees from excessive defoliation.

Pregnant mares abort as a result of accidentally consuming eastern tent caterpillars wandering about on the grass in search of pupation sites. This phenomena cost Kentucky thoroughbred breeders more than $300 million in lost foals in 2001, when eastern tent caterpillars were unusually abundant. Although this situation has not been documented in Mississippi, cautious horse breeders may wish to avoid pasturing pregnant mares near infested wild cherry trees in the spring.

Management

Most years, the damage eastern tent caterpillars cause is not significant, and no control is necessary. Removing and destroying tents and the caterpillars in them can provide control on smaller trees. You can use a hook fashioned from a clothes hanger and taped to a long pole to remove tents from larger trees (be careful around power lines).

Control

Forest tent caterpillars do not build tents. Note the blue stripe and white, keyhole-shaped spots along the back.

Permethrin, Carbaryl, Cyfluthrin, Cyhalothrin, Bts, Spinosad.

In cases where treatment is necessary to control heavy infestations, or to protect small, susceptible trees, apply foliar insecticide sprays. Often, you can treat small, recently planted trees with a hand sprayer. You can use hose-end sprayers designed for treating trees and shrubs for trees up to 20 to 25 feet tall, but few homeowners have the power- spray equipment needed to treat large trees. Treatment of large trees usually must be performed by licensed commercial applicators. Always be aware of the potential for drift onto neighboring properties and other nontarget sites, and take appropriate precautions to avoid drift-related problems. In many urban settings the potential for problems from spray drift may be so great that you won't want to make such treatments.

Forest Tent Caterpillars

Forest tent caterpillars occur on various oaks, sweetgums, black tupelos, maples, elms, and other hardwood species.

Mature caterpillars are about 2 inches long. Forest tent caterpillars are closely related to eastern tent caterpillars, which they resemble. Like eastern tent caterpillars, the background color of the body is black, etched with gold or yellow markings, and the caterpillars are sparsely covered with long, tan hairs. Forest tent caterpillars have a row of white, keyhole-shaped spots down the center of the back, rather than the solid white line seen in eastern tent caterpillars, and a nearly continuous slate blue line down either side of the body. Despite their name, forest tent caterpillars do not build tents. Instead, they form silken mats on the trunk and larger limbs. There is only one generation per year, and this insect overwinters in the egg stage. Eggs are deposited in dark-colored, somewhat slick masses that encircle smaller twigs. Eggs hatch in early spring as leaves begin to develop.

Damage

Occasionally, forest tent caterpillars occur in outbreak numbers and cause widespread defoliation of forest and landscape hardwoods. During non-outbreak years, individual trees can suffer heavy defoliation. Although this injury is unsightly, trees normally produce a second flush of leaf growth, and hardwood trees usually recover from one defoliation with little long-term adverse effect. But repeated defoliation can reduce tree growth rate.

Management

Protect trees that suffer severe defoliation from further stress, such as a second defoliation or drought stress, for the rest of the season.

Control

Permethrin, Carbaryl, Cyfluthrin, Cyhalothrin, Bts, Spinosad.

In cases where insecticides are needed to control heavy infestations or to protect small, susceptible trees, apply a foliar insecticide. Often, you can treat small, recently planted trees with a hand sprayer. You can use hose-end sprayers designed for treating trees and shrubs on trees up to 20 to 25 feet tall, but few homeowners have the power-spray equipment needed to treat large trees. Treatment of large trees usually must be performed by licensed commercial applicators. Always be aware of the potential for drift onto neighboring properties and other nontarget sites, and take appropriate precautions to avoid drift-related problems. In many urban settings, the potential for problems from spray drift may be so great that you won't want to make such treatments.

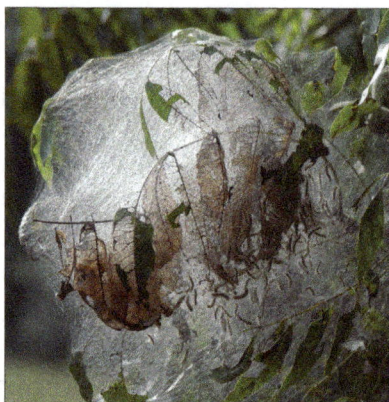

Fall webworms build tents around the ends of branches of pecan and persimmon trees and feed on the leaves enclosed within.

Fall Webworms

Fall webworms primarily attack pecans and persimmons; they occasionally attack other trees, such as Bradford pears.

Fall webworms are important pests of pecans in the home landscape, and heavy infestations can cause severe, and even total, defoliation. Although pecans are commonly grown in the landscape, they are considered a food crop, so you must be careful when selecting insecticides.

Mature larvae are about 1.25 inch long. This caterpillar makes conspicuous webs that enclose the ends of branches. Many dozens of caterpillars occur and feed within one web. The webbing protects the caterpillars from birds, insects, and parasites. The caterpillars themselves may be redheaded with light-colored spots or black-headed with dark spots. Regardless of head color, the caterpillars will be sparsely covered with long, light-colored hairs. Although there are two generations per year, these caterpillars are much more common in late summer and fall.

Damage

The webs this insect makes are unsightly, especially when heavy outbreaks result in dozens, or even hundreds, of webs per tree. This caterpillar causes defoliation by feeding on the leaves enclosed within the web, and heavy infestations can result in total defoliation of susceptible trees. In the absence of other stress, trees normally survive and recover from one-time complete defoliation. But pecans that are severely defoliated may suffer from poorly filled nuts in the year of the defoliation and reduced nut load the following year. Heavy infestations of fall webworms occur more routinely in the southern portion of the state.

Management

The "do nothing" approach is the most commonly used option for managing fall webworms. Avoid planting susceptible species (pecans and persimmons). When only a few webs are present on small- to medium-sized trees, you can remove them by using a hook made from a coat hanger and taped to the end of a long pole (be careful around power lines). Destroying the web this way also exposes the caterpillars to predation and parasitism.

Control

Permethrin, Carbaryl, Cyfluthrin, Cyhalothrin, Bts, Spinosad.

Control webworms with contact insecticide sprays that penetrate the webbing. Often, you can treat small, recently planted trees with a hand sprayer. You can use hose-end sprayers designed for treating trees and shrubs to treat trees up to 20 to 25 feet tall, but few homeowners have the power-spray equipment needed to treat large trees. Treatment of large trees usually must be performed by licensed commercial applicators. Always be aware of the potential for drift onto neighboring properties and other nontarget sites, and take appropriate precautions to avoid drift-related problems. In many urban settings, the potential for problems from spray drift may be so great that you won't want to make such treatments. When treating pecans, be sure the insecticide is specifically labeled for pecans (some formulations of carbaryl and malathion are labeled for use on pecans).

Walnut caterpillars occasionally defoliate individual limbs or areas of pecan and walnut trees. Older caterpillars are dark-colored, as seen here, but younger caterpillars are reddish with thin, white, longitudinal stripes.

Walnut Caterpillars

Walnut caterpillars occur on walnut, hickory, and pecan trees.

Although they are gregarious and feed together as a group, walnut caterpillars do not build webs. They have the unusual habit of clustering in masses on the trunk of the infested tree when it is time for them to molt or shed their skin. After molting, the caterpillars move back into the crown of the tree to resume feeding, leaving a mass of shed skins stuck to the trunk of the tree. These masses of shed skins may remain on the tree for some time after the caterpillars have completed their development and crawled to the ground to pupate. The caterpillars are red with white stripes down their sides when young, and black or dark-colored when fully mature. Caterpillars of both colors are sparsely covered with long, white hairs. When they are disturbed, the caterpillars tend to arch their heads and tail ends, creating a wide U-shape. There are two generations per year.

Damage

These caterpillars feed together in large groups, causing localized defoliation within the crown of the tree. Often, all of the leaves on one branch will be eaten, leaving only the larger mid-veins, while leaves of adjacent branches are undamaged. Such localized defoliation seldom causes serious long-term injury.

Management

When masses of caterpillars move to the trunk of the tree to molt, they are often near the ground where you can reach them and physically destroy them, or you can spray them directly with labeled contact insecticides.

Control

Permethrin, Carbaryl, Cyfluthrin, Cyhalothrin, Bts, Spinosad.

If infestations are heavy enough to threaten severe defoliation, you can control caterpillars with foliar insecticide sprays. Small, recently planted trees can often be treated with a hand sprayer. Hose-end sprayers designed for treating trees and shrubs can be used to treat trees up to 20 to 25 feet tall, but few homeowners have the power-spray equipment needed to treat large trees. Treatment of large trees usually must be performed by licensed commercial applicators. Always be aware of the potential for drift onto neighboring properties and other nontarget sites, and take appropriate precautions to avoid drift-related problems. In many urban settings, the potential for problems from spray drift may be so great that you won't want to make such treatments. When treating pecans, be sure the insecticide you are using is specifically labeled for pecans. Some formulations of carbaryl and malathion are labeled for use on pecans.

Bagworms are damaging pests of needle-bearing evergreen trees,
such as junipers, cedars, and arborvitae.

Bagworms

Bagworms most commonly attack arborvitae, junipers, cedars, and other needle-bearing evergreens but also occasionally occur on broadleaf trees and shrubs.

Bagworms are about 1.25 inch long. You can easily recognize this insect by the tapering gray to tan silk bags it produces and attaches to its host plant. Some of the needles or leaves of the plant are usually woven into the bag. If the bags contain caterpillars, they are not firmly attached to the plant, but when the insects pupate, they use strong silk to attach the bag to the plant. Adult females are wingless and never leave the bag. Eggs are deposited inside the bag, and this is how this pest overwinters. In the spring, newly hatched larvae either remain on the original plant or spread to other plants by ballooning on a silken parachute.

Damage

Damage is the result of defoliation caused by the feeding caterpillars. When only a few caterpillars are present, defoliation is negligible, but heavy infestations can result in complete or severe defoliation of individual plants. Defoliated plants covered with bags are unsightly.

Management

Because of this insect's limited mobility, infestations are often localized, and hand-picking before eggs hatch in the spring can effectively control low infestations on small plants. It is easier to remove bags of larvae than pupal cases or egg cases, which are attached to the plant with strong silk. These often have to be cut away with scissors or pruning shears.

Control

Bts, Spinosad.

You can control this pest with foliar insecticide sprays, but choose your treatment carefully. Spider mite populations sometimes increase after applications of carbaryl or pyrethroid insecticides (permethrin, cyfluthrin, and cyhalothrin). Spinosad and Bt products are less likely to flare mites. Treat small, recently planted trees with a hand sprayer, or use a hose-end sprayer designed to treat trees and shrubs up to 20 to 25 feet tall. Treat in mid-April through early June to control newly hatched caterpillars before they cause much damage. Treatments applied in late summer or fall, after caterpillars have pupated, will not be effective.

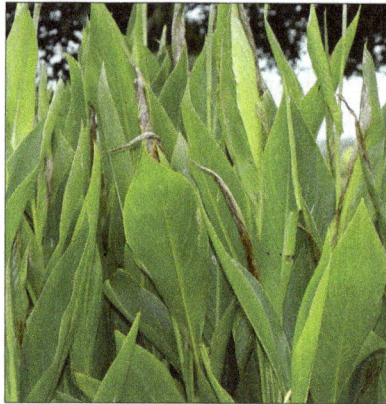

Lesser canna leafroller caterpillars feed within the whorls of cannas,
preventing them from unrolling properly.

Lesser Canna Leafrollers

Lesser canna leafrollers attack cannas.

Mature larvae are about three-fourths inch long. This caterpillar is the most important insect pest of cannas. You can easily identify it by the damage it causes—leaves fail to unroll properly and exhibit a "rat-tailed" appearance. The caterpillars, which are found inside these rolled leaves, have skin that is somewhat translucent and covered with light- colored spots. The larvae look green because you can see the gut contents through the skin. The naked brown pupae are also found inside the rolled leaves. You may see the moths, which are light tan with faint wavy lines of darker brown, resting on the foliage with their wings spread and their antennae folded back along the sides of the body.

Damage

Young larvae begin by feeding as leaf miners, creating small, frass-filled tunnels within the leaf. As the caterpillars grow larger, they leave the leaf mines and bind the young, unrolled leaves with silk, preventing them from unrolling properly. They then complete their development within the protected area of this leaf roll, where they feed on the upper surface of the leaf but leave the lower, translucent epidermis intact, creating a windowpane effect. As many as 10 or more caterpillars may feed inside a single leaf roll. Leaves that do manage to unroll often have mines where young larvae fed, win-dowpaned defoliated areas, and rows of holes across the leaf blade. Heavily infested plantings produce few blooms and are unsightly.

Management

Some gardeners avoid planting cannas because of this pest. Some varieties of cannas are less susceptible than others. These insects overwinter as partly grown larvae and pupae in the whorls and debris of the previous year's foliage. Removing and destroying old stalks and debris in or before late winter can aid greatly in controlling this pest, especially if there are no nearby plantings where old stalks and debris are allowed to remain through the winter. Check cannas regularly for early signs of infestation. If you detect infestations, prune and destroy infested stalks and apply a foliar insecticide.

Control

Acephate, Carbaryl, Bifenthrin, Cyfluthrin, Cyhalothrin, Permethrin, Spinosad.

Control this insect with foliar insecticide sprays. Many effective treatments are available in pre-mixed, ready-to-use formulations, and this can be a quick and convenient way to treat small plantings. Because there are several generations per year, it may take several treatments, applied at intervals throughout the summer, to maintain control. Be sure to direct sprays into the unrolled leaf whorls. Because canna leaves are slick and waxy, it is helpful to add a "sticker" to the insecticide spray. Insecticides that have systemic activity, such as acephate, are often most effective.

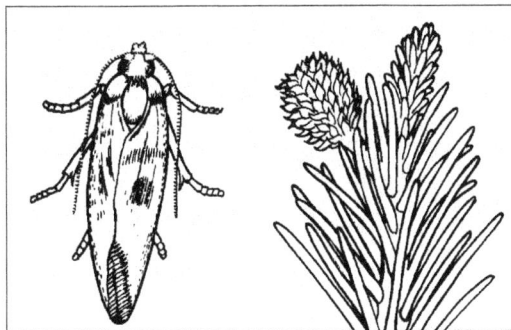

Nantucket Pine Tip Moths

Nantucket pine tip moths attack most pines, including Virginia pines and Scots pines, but not longleaf or eastern white pines; they seldom attack slash pines.

Adult moths are small and seldom seen; they are about one- third inch long and are brick-red mottled with gray. Larvae, which are about one-third inch long when mature, are small, yellow caterpillars with dark heads and cervical shields. The small, shiny, brown pupae are often found inside infested terminals.

Damage

Damage is caused by the larvae, which bore into tips of new growth, causing deformity or death of infested tips. This results in excessive branching and uneven growth on older trees. Attacks to newly planted trees can result in crooked or malformed trunks. Trees less than 10 feet tall are most susceptible. This insect has several generations per year.Emergence of the first generation can occur as early as late February, especially in the southern portion of the state.

Management

Monitor susceptible species of pines and protect them from excessive tip moth damage during the first 3 to 5 years after planting. You can use pheromone traps to monitor for presence of adult moths, but insecticide treatments are required to provide control when damaging insect numbers are present.

Control

Imidacloprid (drench), Acephate, Bifenthrin, Permethrin, Cyfluthrin.

Soil-drench applications of imidacloprid will help control pine tip moths. This is one of the few caterpillar pests that imidacloprid controls. Treatments should be applied in early winter. When properly timed, foliar insecticide treatments control newly hatched larvae and prevent infestation. Several applications per year may be required to properly protect newly planted, susceptible trees.

Redheaded pine sawflies occasionally defoliate young pine trees in home and commercial landscapes.

Sawflies

Sawflies attack pines, ashes, river birches, and various other plants.

Adult sawflies are wasp-like insects, but their biology is very different from the paper wasps with which most people are familiar. During the larval stage, most sawflies feed on the leaves of plants and resemble caterpillars. In fact, sawfly larvae look so much like caterpillars that many people are surprised to learn they are not caterpillars. Sawfly larvae can be distinguished from caterpillars by the fact that they have more than five pairs of prolegs on their abdomens and by the presence of a distinct eyespot on their heads.

Damage

The larvae cause damage by eating the leaves or needles of infested plants, occasionally resulting in severe defoliation. Sawfly populations are very cyclic and sporadic, but heavy infestations occasionally occur on pines, ashes, river birches, and other plants.

Common Species

There are many different species of sawflies. Following are some of the most common:

Redheaded pine sawfly: The larvae of this insect are covered with rows of distinct shiny, black spots on a background of light yellow to green. As the name suggests, the head capsule is reddish, and there is a black eyespot on each side of the head. Outbreaks of redheaded pine sawflies occasionally occur on young pines. Infestations are most common on pines less than 15 feet tall. There are two to three generations per year.

Dusky birch sawfly: This insect is an occasional defoliator of river birch trees. It looks somewhat like the redheaded pine sawfly, having rows of shiny, black spots on a green to yellow background, but it has a dark-colored head capsule. Larvae often rest with the end of their abdomens raised away from the leaf surface in a characteristic S-shaped pose.

Hibiscus sawfly: This is an important pest of certain types of ornamental hibiscus. It is discussed in more detail in the following section.

Control

Permethrin, Bifenthrin, Cyfluthrin, Cyhalothrin, Carbaryl, Malathion, Spinosad.

When you detect them in time, you can control sawfly infestations with sprays of foliar insecticides. Small, recently planted trees can often be treated with a hand sprayer. Hose-end sprayers designed for treating trees and shrubs can be used to treat trees up to 20 to 25 feet tall, but few homeowners have the power-spray equipment needed to treat large trees. Treatment of large trees usually must be performed by licensed commercial applicators. Always be aware of the potential for drift onto neighboring properties and other nontarget sites, and take appropriate precautions to avoid drift-related problems. In many urban settings, the potential for problems from spray drift may be so great that you won't want to make such treatments.

Heavy, lace-like defoliation on the leaves of large-flowered hibiscus/rose mallow plants is usually caused by hibiscus sawfly larvae.

Hibiscus Sawflies

Hibiscus sawflies attack hibiscus, particularly the large- flowered moscheutos varieties.

Hibiscus sawflies are the most important insect pest of the large-flowered moscheutos-type hibiscus. The larvae are small, green, caterpillar-like insects that are only about one- fourth inch long when fully mature. They have black heads and rows of short, raised, spike-like projections along their backs. They feed on the underside of the leaf. Larvae are so small and inconspicuous that the casual observer often overlooks them. Adults are small, wasp-like insects, about one-fourth inch long. The body and head are black, but the thorax is reddish-orange. Eggs are inserted into the leaf tissue near the tip or edge.

Damage

These insects cause heavy defoliation injury to susceptible varieties of hibiscus. Untreated infestations can result in complete defoliation, leaving only the lacy leaf veins. There are several generations per year, and defoliation can occur throughout the growing season.

Management

Native varieties such as grandiflorus and aculeatus are less susceptible to this pest, and many new hybrids exhibit resistance. Check susceptible varieties often throughout the growing season, and treat as soon you detect sawflies. Check for small larvae on the undersides of leaves, for the black and red adults, or for egg-laying wounds on leaves.

Control

Acephate, Permethrin, Bifenthrin, Cyfluthrin, Cyhalothrin, Carbaryl, Malathion, Imidacloprid (drench), Spinosad.

Hibiscus sawflies are easily controlled with foliar insecticide treatments. Because there are several generations per year, several treatments are usually required to provide season-long protection. Many effective treatments are available in pre-mixed, ready-to-use formulations, which makes applying treatments quick and convenient. Systemic insecticides, such as acephate, usually provide longer-lasting control. Soil-drench applications of imidacloprid help control hibiscus sawflies.

Larger elm leaf beetles are infrequent defoliators of elm trees.

Leaf-Feeding Beetles

Leaf-feeding beetles are most commonly found on elms, crape myrtles, willows, and cottonwoods.

Several different species of leaf-feeding beetles occur on landscape plants. These belong to the family of beetles known as Chrysomelidae. This is an unusual group of beetles in that, in many species, both the larvae and adults feed on leaves.

The adults are small beetles approximately one-fourth inch long. Coloration depends on species. Some are colorfully marked; others are metallic blue. Larvae range from black to tan or yellow with black spots or stripes.

Damage

For most species, damage is caused by both adults and larvae, which feed on the leaves and cause defoliation Damaged leaves often have an unsightly skeletonized appearance because of browned, uneaten leaf veins and cross veins.

Common Species

A few of the most commonly encountered species of leaf- feeding beetles are briefly discussed below.

Elm leaf beetle: Adults are yellow with longitudinal black stripes. Larvae are also yellow with dark stripes down each side. Both adults and larvae feed on leaves, and the yellow eggs are deposited in clusters on the undersides of leaves. It attacks all species of elms but is more common on some species than others.

Cottonwood leaf beetle: Adults are yellow with striking long, black marks. Smaller larvae are dark-colored, but larger larvae are gray with black spots. Both adults and larvae feed on leaves, causing skeletonizing defoliation. This insect is found on eastern cottonwoods as well as willows and other species of poplar. There are also several closely related species.

Altica foliacea: This flea beetle has no common name but is sometimes referred to as the "crape myrtle flea beetle." The adults are small, shiny, blue-green metallic-colored insects that jump when disturbed. Adults and larvae feed on weed hosts, occasionally building to high numbers. High numbers of adults can occur on crape myrtles in midsummer, causing damage by injuring the leaves. They are most commonly found in nurseries or on recently planted trees.

Control

Carbaryl, Bifenthrin, Cyfluthrin, Imidacloprid, Cyhalothrin, Permethrin, Malathion, Spinosad.

Leaf beetle populations are often kept in check by predators and parasites. When outbreak populations occur, foliar insecticide sprays are required to minimize damage. Repeated applications may be required to control heavy or persistent infestations. Although not generally effective against beetles, spinosad is active against the larval stage of some leaf-feeding beetles.

Yellow Poplar Weevils

Yellow poplar weevils attack magnolia, yellow poplar, and sassafras trees.

Adults are small (about one-eighth inch long), stout-bodied, gray weevils. The larvae are white, legless grubs that feed inside leaf mines.

Damage

Infestations of concern most commonly occur on Southern magnolias. Adults cause damage by feeding on buds and tender new leaves. Damage is often magnified as leaves expand, resulting in small holes in the leaves. Damage to fully expanded leaves results in numerous brown feeding spots, usually concentrated near the tips of the leaves. Larvae feed as leafminers in large, puffy, blotch-shaped mines at the tips of leaves. Several larvae may occur within one leaf mine. Although infestations are rarely so severe as to adversely affect tree health, heavily infested trees are unsightly. Because of the longevity of magnolia leaves, damage may be evident even when insects are no longer present.

Management

Because this insect overwinters in leaf litter of host trees, raking leaf litter in the fall or early winter can reduce overwintering populations.

Control

There are many species of May beetles. All are robust, brown-colored beetles that fly at night.

Little information is available on controlling this pest, and no insecticides are labeled specifically for yellow poplar weevils. Soil drenches of imidacloprid have shown some promise. Foliar insecticide sprays can be used to control overwintering adults in the spring, as soon as they begin feeding on buds and young leaves, with the objective of controlling the adults before they have a chance to begin laying eggs. Products such as carbaryl, permethrin, or cyfluthrin + imidacloprid should control adult weevils, but timing of sprays is critical, and multiple treatments will be required.

May Beetles

May beetles attack various hardwood trees.

These insects occasionally cause mysterious defoliation of young hardwood trees planted in landscapes.

May beetles are the adult stage of white grubs, which can be important pests of turf grass. They are robust, heavy-bodied, brown to tan beetles about one-half inch long. These beetles are active primarily at night.

Damage

Young oaks and other hardwoods occasionally sustain mysterious defoliation injury that appears to have occurred overnight, yet no insect pests can be found on the tree.

Damaged leaves may be totally consumed, leaving only the petiole and midvein. This type of injury is caused by adult May beetles, which occasionally congregate on individual trees in large numbers during late spring. The beetles feed heavily during the night but leave the tree to seek shelter during the day. This injury is rather uncommon and most often occurs on newly planted trees less than 15 feet tall.

Management

Attacks by May beetles are sporadic and difficult to predict. Young trees near outdoor lights that are left on overnight seem to be most susceptible to attack. Where feasible, turning off overnight lights located near young, susceptible trees during May and early June may reduce the probability of attack.

Control

Permethrin, Bifenthrin, Cyfluthrin, Cyfluthrin + Imidacloprid, Carbaryl.

Although May beetles are susceptible to foliar insecticide treatments, timing of control is difficult because of the unpredictable nature of attack. Foliar sprays may provide some short-term residual control and may help in certain high-risk situations.

Adult Japanese beetles cause damage by feeding on leaves and blooms of many ornamental plants. They can be especially important pests of roses.

Japanese Beetles

Japanese beetles attack the foliage and blooms of many ornamental plants.

The Japanese beetle has progressively expanded its range over the eastern United States since it was introduced in the early 1900s. It has recently become a serious pest of roses and other susceptible plants in North Mississippi and continues to expand its range.

Japanese beetles are about one-half inch long, shiny, metallic green with metallic bronze wings and rows of white, fuzzy spots toward the end of the abdomen. This makes them fairly easy to recognize, although some other native scarab beetles are also metallic green. Larvae are "white grubs" that are about 1 inch long when fully mature. They feed on the roots of grasses in commercial turf, as well as in pastures, roadsides, and unmanaged areas.

Damage

Damage is caused by the adults, which feed on the foliage and blooms of many different species of ornamental plants. This includes many trees, woody shrubs, herbaceous perennials, and annual plants. Leaf-feeding results in skeletonizing defoliation, with the severity of damage being related to the number of beetles present. Japanese beetles are also highly attracted to the blooms of many plants, especially large, light-colored blooms. They feed on the anthers and petals, causing unsightly damage. White or yellow roses are favorite targets and often sustain heavy damage in areas where Japanese beetles are common.

Management

In areas where Japanese beetles are abundant, gardeners may choose to avoid growing species or varieties of plants that are especially susceptible. Be careful in making this determination because susceptibility can vary greatly within a group of plants. For example, Japanese maples are quite susceptible, while red maples are seldom attacked. Likewise, some varieties of crape myrtles are relatively resistant to attack, while others are highly susceptible. Light-flowered varieties of roses are more likely to be attacked than are varieties with darker-colored blooms. Biological treatments, such as the milky spore disease bacteria and Japanese beetle traps, are sold commercially, but their effectiveness is not supported by results from controlled experiments. Because larvae can also develop in pastures and wastelands, controlling larvae in commercial turf may not help prevent attack of landscape plants by large numbers of adults.

Control

Carbaryl, Cyfluthrin + Imidacloprid, Bifenthrin, Permethrin, Cyhalothrin, Cyfluthrin, Azadirachtin.

Adult Japanese beetles can be controlled with foliar sprays, and foliar sprays may also provide short-term residual protection. Repeated treatments will be necessary to protect susceptible plants adequately where this pest is abundant. Repeated applications of products containing azadirachtin—a natural product derived from neem seed— also repel Japanese beetles. Be sure to observe pollinator protection cautions when treating plants that are in bloom.

Wood-boring Insects

Close examination of a dead or dying tree almost always reveals wood-boring insects, but this does not necessarily mean insects are the primary cause of the tree's distress. Hundreds of different species of insects are attracted to dead or dying trees, and many families of beetles and moths specialize as wood-borers. In most cases, infestations of wood-boring insects are the result, not the cause, of a tree's distress. Many species of wood-borers attack only dead wood, and many others can successfully initiate an attack on trees that are stressed and in decline. Relatively few species can successfully attack healthy trees. Some of these are discussed specifically in the following sections. Even these species are most likely to attack trees that are stressed or injured.

Healthy, vigorously growing trees defend themselves from attack by wood-boring insects in many ways. Tight, healthy outer bark provides a physical barrier to insect entry. Pines and other species produce large amounts of resin that can "pitch-out" or entomb, attacking insects. Invading larvae can be crushed or walled off by rapid, vigorous growth. Healthy trees also produce many chemical defenses. However, stressed or injured trees produce characteristic odors that are highly attractive to wood-boring insects, signaling that the tree is susceptible to attack.

The most important thing you can do to protect landscape trees from wood-boring insects is to maintain tree health by avoiding injury or stress. Following are some of the key points to keep in mind:

- Choose planting sites carefully, and match the site to the species you are planting.

- Use trunk wraps to protect newly planted trees.

- Avoid tight guy wires or trunk wraps that can injure bark (be sure to remove trunk wraps).

- Use mulch or trunk guards to prevent injury from string trimmers or mowers.

- Avoid using too much mulch, and do not pile mulch against base of trunk.

- Avoid physical injury to trunk or bark.

- Prune properly and at the right time of year.

- Avoid injury to roots and trunk during construction operations.

- Avoid compacting soil over roots.

- Keep trees adequately watered during periods of drought.

- Provide adequate nutrition, but avoid overfertilization.

- Avoid chemical or herbicide injury.

The following sections provide additional information about some of the more important species of wood-borers that will attack healthy, or relatively healthy, landscape trees in Mississippi.

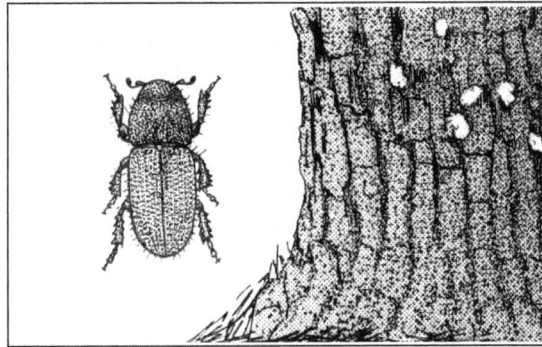

Black Turpentine Beetles

Black turpentine beetles damage only pines.

Adult black turpentine beetles are one-fourth to one-third inch long. These barrel- or cylinder-shaped beetles are reddish-brown to black. The larvae are small, legless, white grubs.

Damage

This insect bores through the outer bark of pine trees and lays its eggs in the inner bark, or cambium layer. The developing larvae feed as a group, creating pocket-like galleries in the inner bark. Multiple attacks can effectively girdle the tree, resulting in death. Popcorn-sized masses of dried resin, known as pitch tubes, on the outer bark often mark the site of an attack. These pitch tubes often contain reddish-brown boring dust.

Attacks by black turpentine beetles are normally restricted to the lower 6 to 10 feet of the trunk. Trees suffering injury or stress are most susceptible to attack. Large, overmature, and, thus, high-value trees, are generally more susceptible than young trees. Attacks are most common from May through September.

Management

Avoid injury or stress to trees. Trees that have been physically injured are more likely to be attacked. Stressed trees are less able to defend themselves by producing resin. Avoid situations that cause soil compaction, root injury, or drought. Keep trees well-watered during periods of drought. Remove dead and dying trees, as well as stumps, as quickly as possible to prevent beetles from breeding in these and then moving to nearby healthy trees.

Control

Permethrin.

When infestations of black turpentine beetles are evident on nearby trees or stumps, it may help to apply preventive insecticide treatments to the trunks of high-value trees. Trunk sprays containing permethrin can help protect trees, and several brands are labeled for homeowners. Be sure to choose a permethrin product that specifically lists use on pine trees to control borers and allows adequate rates of permethrin (0.5% concentration in finished spray). You may need to repeat sprays several times during the summer for adequate protection.

Onyx Pro (bifenthrin) is not labeled for use by homeowners but may be applied in the home landscape by licensed commercial applicators. Because this treatment provides longer residual activity, it may be worth hiring a commercial applicator to treat high-value trees during periods of high beetle populations.

Southern Pine Beetles

Southern pine beetles damage only pines.

Pine beetle adults are about one-eighth inch long. They are smaller than black turpentine beetles but are similarly shaped. They are cylindrical and reddish-brown to dark-colored. The larvae are small, legless grubs.

Damage

Like black turpentine beetles, this insect bores through the outer bark of pine trees and lays its eggs in the inner bark, or cambium layer. The beetles eat winding, S-shaped galleries in the cambium. Adult beetles inoculate infested trees with a fungal disease, which hastens tree death, and larvae also feed in the cambium, girdling this life-supporting tissue.

Trees defend themselves by producing extra resin where the adult beetle tries to bore into the tree. Healthy, unstressed trees can often successfully pitch-out attacking beetles. Because of fungal disease and the girdling of the cambium caused by the winding feeding tunnels, successful invasion by only a few beetles usually kills the tree. Southern pine beetles normally focus their attacks on the main tree trunk, from chest height to where the lower limbs are attached.

Trees suffering injury or stress are most susceptible to attack.

Management

Keeping landscape trees well watered during drought is the most important thing you can do to help reduce the probability of bark beetle attack. Avoid injury or stress to trees. Trees that have been physically injured are more likely to be attacked. Avoid situations that cause soil compaction, root injury, or drought. Remove dead and dying trees, as well as stumps, as quickly as possible to prevent beetles from breeding in these and then moving to nearby healthy trees.

Control

Permethrin.

When infestations of southern pine beetles are active in nearby trees, it may help to apply preventive insecticide treatments to high-value trees. To be effective, though, such treatments must thoroughly cover the trunk from its base to where the first lower limbs are attached. Few homeowners have the necessary equipment to treat larger trees properly. Trunk sprays containing permethrin can help protect trees, and several brands are labeled for use by homeowners. Be sure to choose a permethrin product that specifically lists use on pine trees to control borers and allows adequate rates of permethrin (0.5% concentration in finished spray). You may need to repeat sprays several times in the summer to get adequate protection.

Onyx Pro (bifenthrin) is not labeled for use by homeowners but may be applied in the home landscape by licensed commercial applicators. Because this treatment provides longer residual activity, it may be worth hiring a commercial applicator to treat high-value trees during periods of high beetle populations.

Ips Beetles

Pines are the most common trees attacked by Ips beetles, but some species also attack certain hardwoods.

Adult beetles are so small (about one-sixth inch long) that they are rarely seen. There are several species, all of which have a distinctive indention at the end of the abdomen that is bordered by two rows of small spines. The larvae are small, legless grubs.

Damage

The Ips beetle is one of the more common pests of landscape pines. Attacks by Ips beetles are normally focused on the trunk and large limbs located in the upper, or crown, area of the tree. If the tree is already stressed before attack, reddish boring dust can be found in bark crevices. Attacks to trees that are relatively healthy may result in resin flowing from bore holes and even the formation of small pitch tubes, similar to those caused by southern pine beetles. Adult beetles bore through the outer bark and excavate galleries in the inner bark, or cambium layer. The larvae construct their own feeding galleries that branch out from the initial "egg gallery". When the larvae mature and emerge from the tree as adult beetles, they leave numerous small, round emergence holes that cause the bark to look as though it has been riddled with birdshot.

Management

Keep trees well watered during drought. Avoid injury or stress to trees. Trees that have been physically injured are more likely to be attacked. Stressed trees are less able to defend themselves. Avoid situations that cause soil compaction, root injury, or drought. Remove dead and dying trees, as well as stumps, as quickly as possible to prevent beetles from breeding in these and then moving to nearby healthy trees. Ips beetles also breed in freshly cut pine logs and pulpwood, as well as freshly trimmed limbs. Promptly removing or burning such material reduces the potential for attack.

Control

Keeping trees well watered during drought is the most important thing homeowners can do to help reduce the probability of bark beetle attack. Because Ips beetle attacks are concentrated in the crown area, preventive insecticide treatments are difficult to use effectively.

Densely compacted columns of frass protruding from the bark of a tree are usually a sign of infestation by granulate ambrosia beetle.

Granulate Ambrosia Beetles

Granulate ambrosia beetles (Xylosandrus crassiusculus) attack and kill many species of ornamental hardwoods, including Bradford pears, crape myrtles, ornamental cherries, maples, magnolias, sweet gums, and pecans.

Adult granulate ambrosia beetles are about one-tenth inch long. This small beetle is reddish-brown, stout, and cylinder- shaped. It has a somewhat humpbacked appearance, and its head points downward. When trees are first being attacked, curved, toothpick-sized columns of tightly packed frass, extending up to 3 inches long, often extrude from the bore holes. These columns of frass are somewhat characteristic of this species and are eventually broken off by wind and rain, leaving only birdshot-sized, frass-packed entrance holes.

This is a relatively recent invading pest that seems to be increasing in importance.

Damage

Although granulate ambrosia beetles readily attack dead or dying trees, they are a particular threat to young, recently established trees. Such trees are still under stress from planting. Attacks are more common on trees fewer than 3 inches in diameter, and attacks to young saplings are most likely to occur in the very early spring, at or slightly before bud break. Attacks may occur later in the year, as well. Invading beetles inoculate trees with a fungal disease, which serves as food for the larvae. While some trees may survive localized attacks, trees with numerous frass columns or bore holes around the main trunk do not survive.

Management

Promptly cut and destroy heavily infested trees to reduce potential for spread to uninfested trees. Avoid injury or stress, which increase the potential for attack.

Control

Permethrin.

Trunk sprays containing permethrin can help protect trees, and several brands are labeled for homeowner use. Be sure to choose a permethrin product that specifically lists control of borers and allows use of adequate rates of permethrin (0.5% concentration in finished spray). Trunk sprays applied in late winter may help prevent attacks to susceptible, high-value trees.

Flatheaded appletree borers belong to a group of beetles known as shorthorn beetles.

Onyx Pro (bifenthrin) is not labeled for use by homeowners but may be applied in the home landscape by licensed commercial applicators. Because this treatment provides longer residual activity, it may be worth hiring a commercial applicator to treat susceptible, high-value trees during periods of high beetle populations.

Flatheaded Appletree Borers

Flatheaded appletree borers attack many different species of hardwood trees.

Adult flatheaded appletree borers are about one-half inch long; larvae are about 1 inch long. Adults are somewhat oval-shaped, short-horned beetles and are metallic gray. The larvae are legless, white, segmented, and worm-like. They have small, dark-colored heads, and the three white segments immediately behind the head are enlarged and flattened.

Damage

Although this insect is most attracted to trees that are stressed or injured, it will also attack relatively healthy trees, especially young trees just being established. The larvae damage the tree by boring in bark, cambium, and wood. It takes only one or two larvae to kill or seriously injure a young sapling. There is only one generation per year, but egg-laying may occur throughout much of the growing season.

Management

Healthy, vigorously growing trees often can defend themselves against attack. Avoid injury or stress, which can predispose trees to attack. Trunk wraps can be used to protect newly planted trees during the first couple of years after planting. Be sure support wires do not injure the tree, and avoid excessive mulch around the base of the tree and planting too deeply.

Control

Permethrin.

If you detect larvae actively boring in the trunk, you may be able to lessen damage by digging them out with a knife or using a small, flexible wire to probe into the galleries and puncture the larvae. Trunk sprays of permethrin (0.5% concentration in finished spray) or bifenthrin (available to licensed commercial applicators only) may help protect high-value trees.

Dogwood Borers

Dogwood borers attack many species of hardwoods, but dogwoods are the most commonly damaged landscape species.

Adults are small (less than one-half inch long), day-flying, wasp-like moths that are blue/black with yellow bands around the abdomen. The wings are mostly clear with black tips and markings. Fully mature larvae are slightly over one- half inch long and are cream-colored with reddish-brown heads.

Damage

Eggs are deposited on the outer bark, and newly hatched larvae enter the tree through wounds and small cracks in the bark. The larvae cause damage by boring in the inner bark layer of the tree. Loose, scaling bark or swollen, knotty areas on the trunk indicate infestation. Heavily infested trees may suffer from limb or crown dieback and poor growth.

Management

Keeping trees healthy and vigorous and avoiding injury and stress are the most important means of preventing infestation. Choose planting sites carefully. Trees planted in full sun are more susceptible to attack than trees planted in partial shade. On trees that are exposed to full sun, avoid pruning lower limbs to allow more shading of the trunk. Although no cultivars of dogwood are resistant to this pest, some types (such as Korean dogwood) are less susceptible.

Control

Permethrin.

You can use trunk sprays of permethrin to partially protect heavily infested or high-value trees. Sprays will not control larvae that have already bored into the tree. The objective is to have a residue of insecticide on the bark to control newly hatched larvae before they bore under the bark. Sprays must be applied several times during the growing season to obtain season-long control. Egg-laying moths may occur from spring through September, but heaviest populations occur later in the season.

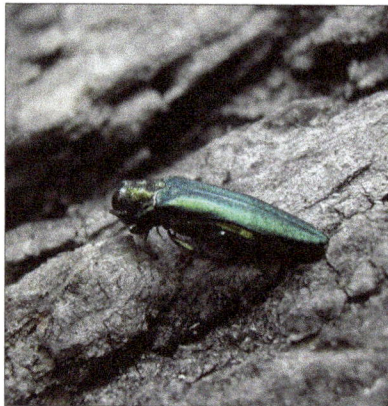

Adult emerald ash borers are strikingly colored, but are only about one-third inch long. Adults are often not noticed until nearby ash trees have already been seriously damaged.

Emerald Ash Borers

As of the fall of 2017, emerald ash borers have not yet been detected in Mississippi, but they are present in all neighboring states and could appear here at any time. This is a serious pest of ash trees that has already killed most ash trees in the eastern United States. Effective treatments are available for protecting high-value landscape trees from emerald ash borers, but these treatments must be applied preventively.

Adults are about one-third inch long; mature larvae are 1 to 1.25 inch long. Adults are small, peg-shaped, metallic-green beetles. The larvae, which live beneath the outer bark, have slender, white, segmented bodies, with many of the segments appearing bell-shaped. Infestations are usually first detected by observation of damage symptoms, including dieback of limbs in the crown of the tree; splitting of bark; presence of small, D-shaped emergence holes (about one-eighth inch in diameter); new growth sprouting from the lower trunk of the tree; and woodpecker feeding. There is only one generation per year. Adults lay eggs in bark crevices in the spring, and larvae feed beneath the outer bark through the summer and fall, overwinter, and emerge as adults the following year.

Damage

Emerald ash borers attack only ash trees. Damage is caused by the larvae, which bore in the cambium, or inner bark, girdling tree limbs and trunks. Damage usually progresses down from the crown of the tree. Trees are usually killed after 3 to 5 years of infestation. Experience in other states indicates that trees that have lost more than half of their canopy due to crown dieback usually cannot be saved. Mortality of untreated trees is around 99 percent.

Management

Preventing initial introduction is the first line of defense. Avoid transporting firewood or ash logs into the state. Trees that become infested with emerald ash borers, and that will not be treated, should be immediately cut and destroyed in order to prevent emergence and spread of adult beetles. Avoid using ash trees in new landscape plantings. Choose alternative species that are not susceptible to emerald ash borers.

Control

To protect high-value landscape ash trees, be alert for reports of the presence of emerald ash borer infestations within 15 to 30 miles of your location, and begin preventive insecticide treatments promptly. Depending on choice of treatment, plan on treating yearly or every second year for as long as you want to protect the tree. Optimum treatment timing also depends on choice of insecticide and method of application; timing ranges from early to late spring (around March 1 through mid-May), with slow-acting, soil-applied treatments needing to be applied earlier than trunk- injection treatments. However, treatments can and should be applied at other times (spring through fall) if you learn of nearby infestations of emerald ash borers.

Systemic insecticides that move through the tree to kill young larvae feeding in the cambium are the most effective treatments for emerald ash borers. Such treatments are applied in one of three ways: as soil drenches or soil injections around the base of the tree, as injections into the trunk of the tree, or as systemic trunk sprays. Soil drenches are most appropriate for homeowners. Trunk injections and trunk sprays are best applied by licensed commercial applicators, and commercial applicators can also use treatments that are applied as soil drenches or soil injections.

Soil-applied treatments available for use by homeowners and commercial applicators include imidacloprid or dinotefuran. These treatments must be reapplied annually in order to provide effective protection. Soil-applied treatments are somewhat less effective and consistent than some of the more effective trunk-injection treatments. For trees larger than 15 inches diameter at breast height, it is best to have trees professionally treated with an effective trunk- injection treatment.

Trunk-injection treatments that can be applied by licensed commercial applicators include imidacloprid or emamectin benzoate. Although more costly than other treatments, emamectin benzoate (Tree-age) is one of the more effective treatments, and a single application will provide control for 2 to 3 consecutive years. Thus, trees treated with emamectin benzoate must be re-treated every second year.

Systemic trunk sprays that contain dinotefuran (Safari 20 SG) can also be applied by commercial applicators. These treatments must be repeated annually.

Adult twig-girdlers chew through pencil-sized twigs of hardwood trees, causing them to break off and fall to the ground. The larvae develop inside these fallen twigs.

Twig-Girdlers

Twig-girdlers usually attack pecan or hickory trees but sometimes occur on other hardwoods.

Twig-girdlers are large (about 1 inch long), gray-brown, long-horned beetles.

Damage

This insect does not attack the trunk of the tree. The female girdles pencil-sized twigs by chewing away a ring of wood and deposits her eggs in the girdled twig. These girdled twigs eventually fall to the ground, and the larvae complete their development inside the fallen twigs. Excessive damage can disfigure young trees and slow growth.

Control

Gather and burn fallen twigs, or place in plastic garbage bags and dispose of in the garbage. This prevents larvae from developing and reinfesting trees. No chemical control is recommended.

Marble oak galls and oak apple galls are two common insect-induced galls of oak leaves.

Gall-forming Insects

Gall-forming insects attack oaks, dogwoods, maples, sugarberries, and many other species.

Many insects can cause galls or unusual growths on ornamental plants. Many of these are small wasps or tiny flies, but several other types of gall-forming insects and some species of mites also cause galls. Gall-forming insects are usually host-specific. The plant makes galls in response to chemicals released by the insect, resulting in the growth of plant tissue around the insect. Many disease organisms also cause galls, and careful identification is necessary to distinguish between galls caused by insects and galls caused by pathogens. Examples of a few of the more common insect-induced galls are discussed here.

Pecan phylloxera: See page 15 for information on this gall- forming insect.

Gouty oak galls: These are the large (golf ball-sized), knot- like growths on the twigs of oaks. A small, cynipid wasp causes these galls. There are two generations per year; the first causes small, blister-like galls on the leaves, and the second generation causes the knotty stem galls. Horned oak gall is a similar gall caused by a related species.

Oak apple galls: Oak apple galls are round, marble- to golf- ball-sized, spongy galls on the undersides of the leaves of various red oak species. They are caused by a small, cynipid wasp. Each gall contains only one larva that develops within a smaller, seed-like capsule in the center of the gall.

Dogwood club gall midges: These are the larvae of a small fly that deposits its eggs in very small, developing dogwood leaves at the tips of the terminal. The resulting larvae, or maggots, burrow into the tip of the developing shoot, causing the end of the twig to become enlarged and club- like. Several dozen of the small, orange maggots may be inside one club gall. Heavy infestations can result in stunted and malformed trees.

Yaupon psyllid galls: This small, aphid-like insect causes the leaves of yaupon hollies to become distorted, creating a folded, pouch-like leaf gall. Several developing nymphs may be inside one gall.

Erineum galls: Red or green, very small mites belonging to the group known as erio-phyid mites cause these felt-like patches on the undersides of maple leaves. Other species of eriophyid mites cause various types of leaf-distortion galls, and there are many species of eriophyids that cause bud- proliferation galls and "witch's broom" galls.

Management

Pruning and destroying galls before the developing insects have time to emerge is an effective control method on small plants, but this is obviously not feasible for large trees. Although galls may be unsightly, they rarely cause severe economic damage, and chemical control is generally not recommended.

Control

Insecticides are not practical or effective for most gall- forming insects. Soil drenches containing imidacloprid may be effective against certain species of gall-forming sucking pests, such as yaupon psyllid galls. Foliar sprays can reduce gall incidence, but timing is very critical, since there is only a very narrow window of time when susceptible adults are present.

Beneficial Insect

Beneficial insects (sometimes called beneficial bugs) are any of a number of species of insects that perform valued services like pollination and pest control. The concept of *beneficial* is subjective and only arises in light of desired outcomes from a human perspective. In farming and agriculture, where the goal is to raise selected crops, insects that hinder the production process are classified as pests, while insects that assist production are considered beneficial. In horticulture and gardening; pest control, habitat integration, and 'natural vitality' aesthetics are the desired outcome with beneficial insects.

Encouraging beneficial insects, by providing suitable living conditions, is a pest control strategy, often used in organic farming, organic gardening or integrated pest management. Companies specializing in biological pest control sell many types of beneficial insects, particularly for use in enclosed areas, like greenhouses.

Types

Some species of bee are beneficial as pollinators, although generally only efficient at pollinating plants from the same area of origin, facilitating propagation and fruit production for many plants. Also, some bees are predatory or parasitic, killing pest insects. This group includes not only honeybees, but also many other kinds that are more efficient at pollinating. Bees can be attracted by many companion plants, especially bee balm and pineapple sage for honeybees, or Apiaceae like Queen Anne's lace and parsley, for predatory bees.

Bee boxes at an organic farm.

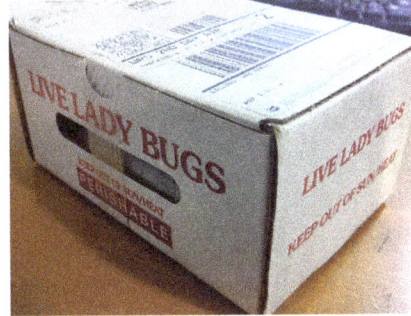

Ladybirds are a beneficial insect commonly sold for biological control of aphids.

A European mantis hunts for prey on a shrub rose.

Encarsia formosa, an endoparasitic wasp, was one of the first biological control agents developed.

Ladybugs are generally thought of as beneficial because they eat large quantities of aphids, mites and other arthropods that feed on various plants.

Other insects commonly identified as beneficial include:

- Assassin bug.

- Damsel bug.

- Earwig.

- Green lacewing.

- Ichneumon wasp.

- Lady bugs.

- Mealybug destroyer.

- Minute pirate bug.

- Soldier beetle.
- Syrphid fly.
- Tachinid fly.
- Trichogramma wasp.

Agricultural Pest Nematodes

Nematodes are common soil pests that affect plants. The aboveground symptoms of disease caused by nematodes can be difficult to detect, and may be often confused with symptoms of nutrient deficiency. Typically, plants do not thrive, are paler than normal, and may wilt in the heat of the day. Affected plants are often dwarfed, with small leaves. Sometimes, when infected plants are growing in moist, fertile soil, or during cool weather, the aboveground parts can still appear healthy.

Heterodera Carotae

Heterodera carotae is a plant pathogenic nematode commonly known as the carrot root nematode or carrot cyst nematode. It is found in Europe, Cyprus and India and is considered an invasive species in the United States. It causes damage to carrot crops and is very specific in its choice of hosts, only infecting Daucus carota and Daucus pulcherrima.

Distribution

In Europe this nematode has been recorded in Germany, France, Portugal, Italy, Hungary, Poland, Sweden, Russia, the Czech Republic, the Netherlands, the United Kingdom and Ireland. It has also been found in India and Cyprus and at several locations in the state of Michigan in the United States of America.

Morphology

The female carrot cyst nematode is white and lemon-shaped, averaging 400 µm by 300 µm, with a pair of ovaries occupying most of the body cavity. The male is thread-like, short with a rounded tail and a single testis averaging 60% of the body length. The mature cyst is lemon-shaped, white at first later becoming reddish-brown, with a distinct neck.

Life Cycle

Eggs are found in the cysts attached to the root systems of carrot plants and in plant debris and contaminated soil. Some hatch soon after the cyst is formed and the second stage juveniles disperse through the soil and invade young rootlets by piercing through the epidermis with their stylets. Most, however, remain in the cyst for two to three months after it has turned brown. At first the male and female juveniles look similar, both being threadlike and growing to 1.5 millimetres long. Four weeks after invasion the juveniles moult and the females become mature. An egg-sac begins to form and soon fills up with from 200 to 600 eggs. As the sac swells it bursts through the root tissues. The males move through the soil searching for females, and after fertilisation, the juveniles begin to develop inside the female's body. She dies and her cuticle hardens to become a cyst still attached to the root.

Economic Significance

Symptoms of infestation with this nematode include patches of the crop with reduced growth, stunted individual plants with bronzed leaves, small distorted roots, a tangled overgrowth of rootlets and the characteristic cysts. In light soils and when uncontrolled, crop losses caused by this nematode have ranged from twenty to eighty percent. The nematode can be dispersed by the transfer of contaminated soil, plant material and machinery. The dehydrated cysts remain infective in the soil or adhering to roots for up to ten years. If successive carrot crops are grown on the same site, nematode numbers can increase tenfold each year.

Anguina Agrostis

Anguina agrostis (Bentgrass nematode, seed-gall nematode) is a plant pathogenic nematode.

Anguina agrostis was one of the first plant parasitic nematodes to be taxonomically described by J.G. Steinbuch in 1799. While on a "botanical walk", Steinbuch collected samples from a grass that resembled *Agrostis silvatica*. He examined the samples and discovered that the grass was not *A. silvatica* but was rather a degenerate form of *Agrostis capillaris*; he further discovered that *A. silvatica* is not a true species (or variety) but that the misclassification of the grass was due to the formation of galls by *A. agrostis* parasitism. Galls caused by *A. agrostis* have glumes that are 4-5 times longer than normal and can cause yield losses of up to 40-70%. In addition to crop loss, *A. agrostis* associates with pathogenic bacteria *Rathayibacter rathayi* (formerly *Corynebacterium rathayi*) to cause annual ryegrass toxicity in Australia.

Hosts and Distribution

A. agrostis infects bentgrasses within the genus *Agrostis* as well as annual and perennial ryegrasses (*Lolium* spp.). The nematode can also infect 14 other genera of grasses. *A. agrostis* has been found in Australia, New Zealand, Western Europe, former USSR, Canada, and the United States.

Morphology

The lip region is slightly offset (3-4 µm high) and the nematode has a very short stylet (10 µm). *A. agrostis* has a three-part esophagus. The procorpus is cylindrical with a swelling near its midsection. The metacorpus is ovoid in shape and the isthmus is long and narrow. The postcorpus has three glands and is highly developed, but does not overlap the intestine. On average, infective juveniles (J2) measure 530 µm and dauer juveniles measure 760 µm; the increase in size is due to feeding and the formation of lipid droplets or storage bodies. Females range from 1.5-2.7 mm in length, are curved ventrally, and are swollen. The vulva is located near the posterior end and one ovary is present. Males are smaller (1.1-1.7 mm in length), are not as swollen, and have a small bursae that extends subterminally. One testis is present.

Life Cycle and Reproduction

Infective second stage juveniles (J2) find a young host, migrate to areas of new growth and are carried up with the growing point of the plant. They may feed ectoparasitically until formation of the inflorescence, at which time the J2 invades the ovule, becomes sedentary, and a gall begins to form. Within the gall, nematodes progress through three molts to reach adulthood (J3, J4, male and female adults). Reproduction is amphimictic and females can lay up to 1000 eggs. The first molt occurs in the egg and the nematode hatches as a J2. These juveniles undergo anhydrobiosis and become the dormant dauer larvae to withstand the hot summer heat of Australia. Autumn rains rehydrate the dauer juveniles which become active to begin the life cycle over again. Only one generation is produced per year.

Host-parasite Relationship

Infective second stage juveniles colonize plants during the vegetative growth stage and may feed ectoparasitically during this time. When the inflorescence begins to form, the J2s invade the flower ovule and begin to feed endoparasitically. Nematode feeding on floret primordia induces rapid cell division, cell enlargement, and subsequent cell degeneration and collapse. The continuation of this process results in the formation of a large central cavity (in which the now-sedentary nematodes reside) enveloped by a gall wall. Gall size increases rapidly as nematodes grow and reproduce. The gall wall is several cell layers thick. Inner cells of the gall wall (near the cavity) have dense cytoplasm with several mitochondria, indicating high levels of metabolic activity. These cells most likely provide nutrients to the nematodes. The outer layers of the gall wall are unmodified, thereby forming and maintaining the gall structure. As the plant senesces, the galls desiccate and the nematodes undergo anhydrobiosis.

Management

To mitigate the effects of annual ryegrass toxicity, farmers can move their livestock another, uninfected pasture at first sign of toxicity. Hot water treatments or chemical seed treatments have been used to produce clean seed. Infected pastures can be managed by mowing, herbicide treatments, or burning. These techniques eliminate the development of inflorescences and halt the life cycle of the nematode since they only mature to adulthood within the seed gall. *A. agrostis* cannot survive in the soil for more than one year and thus practices such as crop rotation or fallow have proven to be effective in managing the nematode.

Soybean Cyst Nematode

The soybean cyst nematode (SCN), Heterodera glycines, is a plant-parasitic nematode and a devastating pest of the soybean (*Glycine max*) worldwide. The nematode infects the roots of soybean, and the female nematode eventually becomes a cyst. Infection causes various symptoms that may include chlorosis of the leaves and stems, root necrosis, loss in seed yield and suppression of root and shoot growth. SCN has threatened the U.S. crop since the 1950s, reducing returns to soybean producers by $500 million each year and reducing yields by as much as 75 percent. It is also a significant problem in the soybean growing areas of South America and Asia.

Biology

Segment of soybean root infected with soybean cyst nematode. Signs of infection are white to brown cysts filled with eggs that are attached to root surfaces.

The second-stage juvenile, or J2, nematode is the only life stage that can penetrate roots. (The first-stage juvenile occurs in the egg, and third- and fourth-stages occur in the roots). The J2 enters the root moving through the plant cells to the vascular tissue where it feeds. The J2 induces cell division in the root to form specialized feeding sites. As the nematode feeds, it swells. The female swells so much that her posterior end bursts out of the root and she becomes visible to the naked eye. In contrast, the adult male regains a wormlike shape, and he leaves the root in order to find and fertilize the large females. The female continues to feed as she lays 200 to 400 eggs in a yellow gelatinous matrix, forming an egg sac which remains inside her. She then dies and her cuticle hardens forming a cyst. The eggs may hatch when conditions in the soil are favorable, the larvae developing inside the cyst and the biological cycle repeating itself. There are usually three generations in the year. In the autumn or in unfavorable conditions, the cysts containing dormant larvae may remain intact in the soil for several years. Although soybean is the primary host of SCN, other legumes can also serve as hosts.

Pathology

The aboveground symptoms of SCN infection are not unique to SCN infection, and could be confused with nutrient deficiency, particularly iron deficiency, stress from drought, herbicide injury or another disease. The first signs of infection are groups of plants with yellowing leaves that have stunted growth. The pathogen may also be difficult to detect on the roots, since stunted roots are also a common symptom of stress or plant disease. Observation of adult females and cysts on the roots is the only accurate way to detect and diagnose SCN infection in the field.

Locations

- Africa: Egypt.

- Asia: Iran (Golestan Province and Mazandaran Province), China (Hebei, Hubei, Heilongjiang, Henan, Jiangsu, Liaoning), Indonesia (Java), Korean peninsula, Japan, Taiwan (unconfirmed), Russia (Amur District in the Far East).

- North America: Canada (Ontario), USA (Alabama, Arkansas, Delaware, Florida, Georgia, Illinois, Indiana, Iowa, Kansas, Kentucky, Louisiana, Maryland, Minnesota, Michigan, Mississippi, Missouri, Nebraska, New Jersey, North Carolina, North Dakota, Ohio, Oklahoma, Pennsylvania, South Carolina, South Dakota, Tennessee, Texas, Virginia and Wisconsin).

- South America: Argentina (unconfirmed), Brazil (unconfirmed), Chile, Colombia, Ecuador.

Control

Cultural practices, such as crop rotation and the use of resistant cultivars, are used to limit the damage due to SCN. Because SCN is an obligate parasite (requires a living host), a crop rotation involving non-host plants can decrease the population of SCN and has been shown to be an effective management tool. Plants that are already stressed are more susceptible to infection, so good cultural practices, like maintaining soil fertility, pH and moisture can reduce the severity of infection. Chemical control with nematicides is not normally used because the economic and environmental costs are prohibitive.

Agricultural Pest Mites

Mites are minute (usually less than 1mm) arachnids with eight legs when adults. They are often pests of animals and plants, infest stored food products and in some cases transmit diseases.

Mites are a risk for Western Australian primary producers as they impact upon market access and agricultural production.

WA is free from some of the world's major agricultural mite pests. Biosecurity and management measures on your property are vital in preventing the spread of mites.

Brevipalpus Phoenicis

Brevipalpus phoenicis, also known as the false spider mite, red and black flat mite, and in Australia as the passionvine mite, is a species of mite in the family Tenuipalpidae. This species occurs globally, and is a serious pest to such crops as citrus, tea, papaya, guava and coffee, and can heavily damage numerous other crops.

Adults

Adult specimens can grow to 280 μm (0.011 in) long (including the rostrum) and 150 μm (0.0059 in) wide. They are flat, oval, and have a dark green to red-orange colour. The adult males are more wedge-shaped than females. This species has two pairs of legs that extend forward and two extending back. It has two sensory rods on tarsus II that distinguish it from another mite species that is known to occur on the same plants, the privet mite, (*Brevipalpus obovatus* Donn.). A black, "H"-shaped mark occurs on females when raised in temperatures of 68 °F to 77 °F, although this marking is not present at 86 °F.

Tarsus

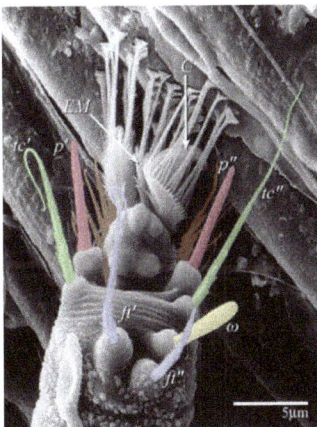

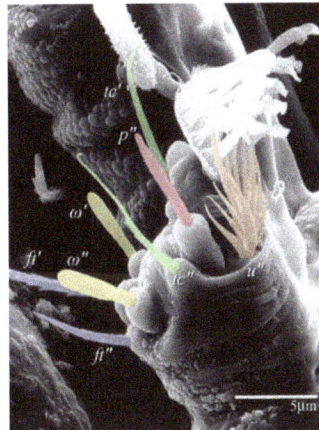

Legend: Empodium (EM) Claw (C) Leg setae-prorals (p) Tectals (tc) Festigials (ft) Unguinals (u) Solenidion (w).

Larvae

Larvae are about 140 µm (0.0055 in) long, have six legs, and are bright orange-red when newly emerged, later becoming opaque-orange. The protonymphs and deutonymphs are somewhat transparent, with some of their inner organs appearing a diffuse green colour, with black or yellow patches. Like the adults, they are eight-legged.

Eggs

Eggs can be seen with the unaided eye, as clusters of reddish-orange.

Distribution

Brevipalpus phoenicis occurs globally, mostly in the tropics. It is known to occur in:

- Argentina,
- Australia,
- Brazil,
- Guyana,
- Cuba,
- Egypt,
- Netherlands,
- India,
- Jamaica,
- Kenya,
- Malaya,
- Mauritius,
- Mexico,
- Spain,
- Taiwan,
- United States.

This species has become established in numerous southern states, throughout the mainland from Florida to California, and also in Hawaii.

Life Cycle

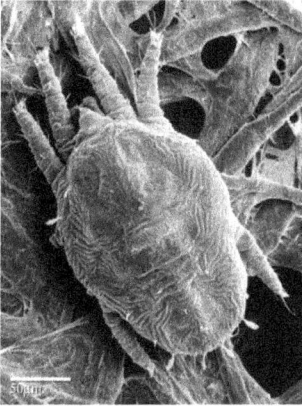
Larva.

Protonymph.

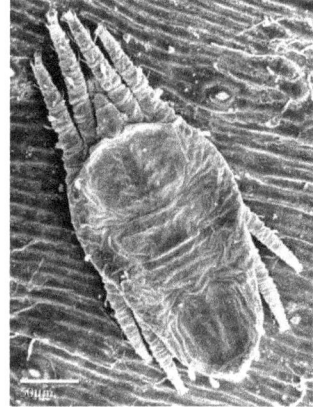
Deutonymph.

Brevipalpus phoenicis lays approximately 50 to 60 eggs during adulthood. These eggs have a fragile stipe that projects from them, and may break if handled. The eggs hatch 8 to 16 days after being laid.

Females deposit eggs singly, commonly sharing a single location with other females. Usually 4 to 8 clusters of eggs are present, normally deposited in cracks or the hollow cavities in leaves created when the internal mesophyll has been destroyed. One day before the eggs hatch, they turn opaque white and the red eyes of the larvae within become visible.

In warm to temperate regions, 4 to 6 generations of this mite can occur each year. In tropical regions, at least 10 generations can occur. Ideal conditions for this species are a temperature range of 25–30 °C (77–86 °F) with high relative humidity.

Adult females live for a maximum of 5 to 6 weeks. The maximum life expectancy for this species is 47 days at 68 °F, with a minimum of 7.5 days at 86 °F in regions of relative humidity of 85% to 90%.

Populations of *Brevipalpus phoenicis* are almost entirely female. This is because the species is parthenogenetic, with most reproduction occurring from unfertilized eggs that produce only females. Development takes place in three stages—larva, protonymph, and deutonymph. The maturation from egg to adult occurs during 12 to 24 days.

Moulting

This mite experiences a quiescent period prior to each moulting, during which it does not eat. During this time it remains attached to the host plant by its stylet alone, with its legs held straight.

Hosts

This species is known to have at least 65 hosts, and the USDA reports that there may be up to 1,000. In the state of Florida, this mite is known to infest Aphelandra, gardenia, grapefruit, hibiscus, holly, ligustrum, lemon, lime, orange, pecan and viburnum.

Damage

This species damages fruit by injecting the cells with toxic saliva. They do this to be able to digest the contents. They puncture numerous cells in close proximity to one another, causing visible chlorotic spots around the area. Later, these spots merge to become brown patches. This can stop the plant's growth and result in deformations. This may cause the skin of the fruit to rupture, and shoots to lose leaves and occasionally die back.

On papaya, the mites begin by feeding on the trunk of the tree. As the population becomes more dense, they migrate to the leaves and fruit. Characteristic evidence of feeding sites is drying of the surrounding areas, with brown colour appearing and the area becoming callous and suberized.

Damage to citrus is more severe. When cholorotic spots are great in number, production of the host plants may be severely reduced. Characteristic galls at the nodes may be observed, and the buds may be unable to sprout. Shoots may be grossly misshapen, and very few leaves may develop. This can result in the death of the entire tree.

Apart from the physical damage this species can cause, it is also a vector of both citrus leprosis and the coffee ringspot virus.

Control

Early in the 20th century, Florida farmers used sulfur as a way to control this pest. However, this is toxic to other, beneficial arthropods.

There are at least four natural predators of this species, but are generally not useful economically, as they attack *Brevipalpus phoenicis* only after the population has increased to very high numbers and severe crop damage has already been done.

Currently, as there is no alternative available, pesticides are used.

Abacarus Hystrix

Abacarus hystrix (cereal rust mite or grain rust mite) belongs to the family Eriophyidae. They are extremely small with adults measuring up to one millimeter in length and only have four legs at the front of the body. Viewing by the human eye requires, a 10 – 20X lens. The adult mites are usually yellow but also have been seen to be white or orange. The cereal rust mite was first found on Elymus repens (couch grass), a very common perennial grass species. It has now been found on more than 60 grass species including oats, barley, wheat and ryegrass, found in Europe, North America, South Africa and Australia. Mites migrate primarily through wind movement and are usually found on the highest basal sections of the top two leaf blades. Abacarus hystrix produces up to twenty overlapping generations per year in South Australian perennial pastures, indicating that the species breeds quite rapidly. It has been noted that the cereal rust mite can cause losses in yield of up to 30-70%.

Life Cycle

Cereal rust mite eggs are exceptionally small and are placed in leaf vein grooves by the mite. The eggs usually begin hatching at the beginning of spring (March in the Northern Hemisphere and September in the Southern Hemisphere) and once they have reached the juvenile stage, the mites mature very quickly (16–18 days). Once the mites are at the adult stage they often travel to the lower section of the plant where they feed on young tissues. Mites are always present for the full growing season of the plant, but activity has been seen to decrease as the temperature begins to rise, this is because unlike other mite species the cereal rust mite favors much cooler temperatures.

Impacts of Mite on Grasses

Abacarus hystrix is a vector for two viruses (Agropyron mosaic and Ryegrass mosaic) and also causes direct damage to the leaf. The effect of the Ryegrass Mosaic Virus (RMV) which is only transmitted by this mite, is chlorotic streaks on the leaves. When the mite feeds on grooves of the leaf surface, it prefers the large cells on the smooth bottom of the groove as opposed to the more ridged, small cells of the side walls. Mite feeding causes direct damage to the leaves, which can be noticed as discoloration or "rusting" of the leaf.

Eradication and Management Options

As a precaution, fields should be checked regularly for mites before spring. By the use of a quadrat system, random plants are selected from different locations in the field. When checking, look for eggs and juvenile mites in the specific area of the leaf veins. A potential management option is to reduce the length of the grass in the cooler months. Studies have shown that trimming grasses reduces the number of mites and since the mites are vectors of many viruses, these viruses are spread less quickly.

Rodents

Rodent refers to any of more than 2,050 living species of mammals characterized by upper and lower pairs of ever-growing rootless incisor teeth. Rodents are the largest group of mammals, constituting almost half the class Mammalia's approximately 4,660 species. They are indigenous to every land area except Antarctica, New Zealand, and a few Arctic and other oceanic islands, although some species have been introduced even to those places through their association with humans. This huge order of animals encompasses 27 separate families, including not only the "true" rats and mice (family Muridae) but also such diverse groups as porcupines, beavers, squirrels, marmots, pocket gophers, and chinchillas.

General Features

All rodents possess constantly growing rootless incisors that have a hard enamel layer on the front of each tooth and softer dentine behind. The differential wear from gnawing creates perpetually sharp chisel edges. Rodents' absence of other incisors and canine teeth results in a gap, or diastema, between incisors and cheek teeth, which number from 22 (5 on each side of the upper and lower jaws) to 4, may be rooted or rootless and ever-growing, and may be low- or high-crowned. The nature of the jaw articulation ensures that incisors do not meet when food is chewed and that upper and lower cheek teeth (premolars and molars) do not make contact while the animal gnaws. Powerful and intricately divided masseter muscles, attached to jaw and skull in different arrangements, provide most of the power for chewing and gnawing.

Capybara (Hydrochoerus hydrochaeris).

The range in body size between the mouse (18 grams [0.64 ounce], body 12 cm [4.7 inches] long) and the marmot (3,000 grams, body 50 cm long) spans the majority of living rodents, but the extremes are remarkable. One of the smallest is Delany's swamp mouse (Delanymys brooksi), associated with bamboo in the marshes and mountain forests in Africa. It weighs 5 to 7 grams, and the body is 5 to 6 cm long. The largest is the capybara (Hydrochoerus hydrochaeris) of Central and South America, which weighs 35 to 66 kg (77 to 146 pounds) and stands 50 to 60 cm at the shoulder, with a body 100 to 135 cm long. Some extinct species were even larger, attaining the size of a black bear or small rhinoceros. The largest rodent ever recorded, Josephoartigasia monesi, lived some two to four million years ago, during the Pleistocene and Pliocene epochs; by some estimates it grew to a length of about 3 metres (10 feet) and weighed nearly 1,000 kg.

Rodents have lived on the planet for at least 56 million years and modern humans for less than one million, but the consequences of their interactions during that short overlap of evolutionary time have been profound. For rodents, early humans were just another predator to avoid, but with Homo sapiens' transition from nomadic hunting and gathering to sedentary agricultural practices, humans became a reliable source of shelter and food for those species having the innate genetic and behavioral abilities to adapt to man-made habitats. The impact of these species upon human populations ranges from inconvenient to deadly. Crops are damaged before harvest; stored food is contaminated by rodent waste; water-impounding structures leak from burrowing; and objects are damaged by gnawing. Certain species are reservoirs for diseases such as plague, murine typhus, scrub typhus, tularemia, rat-bite fever, Rocky Mountain spotted fever, and Lassa fever, among others. Only a few species are serious pests or vectors of disease, but it is these rodents that are most closely associated with people.

Various other rodents are beneficial, providing a source of food through hunting and husbandry, apparel derived from their fur, test animals for biomedical and genetic research (especially mice and rats), pleasure as household pets.

Form and Function

The body form of tree squirrels may be the model for the earliest, and presumably generalized, rodents (genus Paramys). With their ability to adhere to bark with their claws, squirrels adeptly scamper up tree trunks, run along branches, and leap to adjacent trees; but they are equally agile on the ground, and some are capable swimmers. Burrowers are also represented in the form of long-tailed ground squirrels.

Eastern gray squirrel (Sciurus carolinenis).

The specialized body forms of other kinds of rodents tie them closer to particular locomotor patterns and ecologies. Some strictly arboreal species have a prehensile tail; others glide from tree to tree supported by fur-covered membranes between appendages. Highly specialized fossorial (burrowing) rodents, including blind mole rats, blesmols, and ground squirrels, are cylindrical and furry with protruding, strong incisors, small eyes and ears, and large forefeet bearing powerful digging claws. Semi-aquatic rodents such as beavers, muskrats, nutrias, and water rats possess specialized traits allowing them to forage in aquatic habitats yet den in ground burrows. Terrestrial leaping species, such as kangaroo rats, jumping mice, gerbils, and jerboas, have short forelimbs, long and powerful hind limbs and feet, and a long tail used for balance. Body forms of some rodents converge on those in nonrodent orders, resembling shrews, moles, hares, pikas, pigs, or small forest deer. There is also convergence between distantly related groups of rodents in particular body forms and associated natural histories.

Regardless of body form, all rodents share the same basic tools that, as mammologists Emmons and Feer noted, "can be used to cut, pry, slice, gouge, dig, stab, or delicately hold like a pair of tweezers; they can cut grass, open nuts, kill animal prey, dig tunnels and fell large trees".

References

- "Phillip Alampi Beneficial Insect Rearing Laboratory". State of New Jersey Department of Agriculture. Retrieved August 2, 2012

- Pest, definition: maximumyield.com, Retrieved 3 March, 2019

- Berney, M. F., and G. W. Bird. 1992. Distribution of Heterodera carotae and Meloidogyne hapla in Michigan carrot production. Journal of Nematology 24 (4S):776-778

- Pest-insects, pests, pests-weeds-diseases: agric.wa.gov.au, Retrieved 4 April, 2019

- Davis, E.L. and G.L. Tylka. 2000. Soybean cyst nematode disease. The Plant Health Instructor. Doi:10.1094/PHI-I-2000-0725-01

- Rodent, animal: britannica.com, Retrieved 5 May, 2019

- Goodey, T. 1932. The genus Anguillulina Gerv. & v. Ben., 1859, vel Tylenchus Bastian, 1865. Jour Helminthol 10:75-180

3
Weed Control: A Comprehensive Study

Weed control is the method which controls and manages the growth of noxious and invasive weeds. Mechanical weed control, cultural weed control, biological weed control, chemical weed control, integrated weed management, etc. fall under its domain. This chapter delves into these methods of weed control to provide an in-depth understanding of the subject.

Weed control, a botanical component of pest control, stops weeds from reaching a mature stage of growth when they could be harmful to domesticated plants, sometimes livestocks, by using manual techniques including soil cultivation, mulching and herbicides.

Weed control practices in forests are designed to favour the growth of the desired tree species, improve visibility along forest roads, control noxious weeds, and improve wildlife habitats. The goal is to manage timber species, ground vegetation, and wildlife so that each component is maximized yet balanced. Vegetation management is a primary means to achieve a productive forest. Managers need to integrate the best cultural, mechanical, and chemical practices into appropriate and cost effective management systems to minimize losses and detrimental effects due to weeds.

Objectives of Forest Weed Management

A forester might undertake a weed management program with one or more of the following objectives in mind:

Removing unwanted vegetation from planting sites to favor the planted trees. Releasing more desirable species from less desirable overtopping species. Thinning excess plants from a stand. Preventing disease movement through root grafts. Preventing invasion of herbaceous and woody vegetation into recreational areas and wildlife openings. Controlling vegetation along forest roads and around buildings and facilities. Eliminating poisonous plants from recreational areas. Controlling production-limiting weeds in a seed orchard or tree nursery. When establishing a forest, relatively few seeds or seedlings are introduced into an environment in which an almost unlimited number of other plants exist or have the potential to become established. The immediate goal of the forest manager is species survival, which is achieved by reducing the competition from weeds. Site preparation and tree release are the procedures that minimize the density and reduce the vigor of the competing vegetation in the year of and in the years immediately after planting. The type and intensity of management practices depend on the vigor of the desired (planted) species and the indigenous species.

Integrated Weed Management

Integrated Weed Management (IWM) is an approach to managing weeds using multiple control tactics. The purpose of IWM is to include many methods in a growing season to allow producers the best chance to control troublesome weeds.

Weeds negatively impact crop yields, interfere with many crop production practices, and weed seeds can contaminate grain. Based on national research, corn and soybean yield can be reduced by approximately 50% without effective weed control.

Herbicide application is the main weed control strategy used. Reliance on this one method has led to the development of herbicide-resistant weeds. There are a limited number of herbicides available to use and cases of herbicide resistance are rapidly increasing in the US. As a result, herbicides are in need of extra help to continue to ensure adequate weed control.

It is imperative to integrate non-herbicide weed management tactics now to control weeds rather than relying on the ag-chemical industry to continue to develop new herbicides.

Components of an IWM Plan

The goal of IWM is to incorporate different methods of weed management into a combined effort to control weeds. Just as using the same herbicide again and again can lead to resistance, reliance on any one of the methods below over time can reduce its efficacy against weeds. Two major factors to consider when developing an IWM plan are (1) target weed species and (2) time, resources, and capabilities necessary to implement these tactics.

While it is wise to be a good steward of herbicide technology, through the use of PRE and POST herbicide applications or tank mixes, IWM requires the use of tactics beyond herbicides. For example, using these herbicide application practices along with a winter cover crop or harvest weed seed control (HWSC) and prevention methods would be considered IWM.

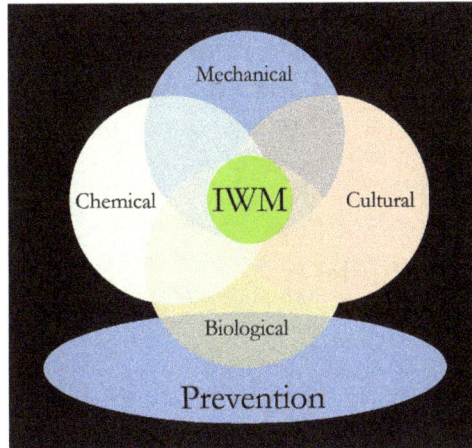

IWM is composed of mechanical, cultural, chemical and biological tactics.

Categories of IWM Practices

Prevention: Prevention is one of the first steps of weed management. This category is unlike the others in that it focuses on keeping weeds out of the field or spreading within a field.

Growers can incorporate this tactic by:

- Avoiding crop seed, manure, and other inputs that are contaminated with weed seeds.

- Cleaning equipment, including combines (combine cleaning methodology), that could transport weed seeds between fields.

- Preventing weeds from producing seeds in the field but also in ditches, fence-rows, and other nearby non-crop areas.

- Scouting for weeds in a timely manner.

- Proceeding with caution when purchasing used farm equipment or using rented land.

Mechanical Weed Control

Mechanical weed control is any physical activity that inhibits unwanted plant growth. Mechanical, or manual, weed control techniques manage weed populations through physical methods that remove, injure, kill, or make the growing conditions unfavorable. Some of these methods cause direct damage to the weeds through complete removal or causing a lethal injury. Other techniques may alter the growing environment by eliminating light, increasing the temperature of the soil, or depriving the plant of carbon dioxide or oxygen. Mechanical control techniques can be either selective or non-selective. A selective method has very little impact on non-target plants where as a non-selective method affects the entire area that is being treated. If mechanical control methods are applied at the optimal time and intensity, some weed species may be controlled or even eradicated.

Mechanical Control Methods

Weed Pulling

Pulling methods uproot and remove the weed from the soil. Weed pulling can be used to control some shrubs, tree saplings, and herbaceous plants. Annuals and tap-rooted weeds tend to be very susceptible to pulling. Many species are able to re-sprout from root segments that are left in the soil. Therefore, the effectiveness of this method is dependent on the removal of as much of the root system as possible. Well established perennial weeds are much less effectively controlled because of the difficulty of removing all of the root system and perennating plant parts. Small herbaceous weeds may be pulled by hand but larger plants may require the use of puller tools like the Weed Wrench or the Root Talon. This technique has a little to no impact on neighboring, non-target plants and has a minimal effect on the growing environment. However, pulling is labor-intensive and time consuming making it a more suitable method to use for small weed infestations.

Mowing

Mowing methods cut or shreds the above ground of the weed and can prevent and reduce seed populations as well as restrict the growth of weeds. Mowing can be a very successful control method for many annual weeds. Mowing is the most effective when it is performed before the weeds are able to set seed because it can reduce the number of flower stalks and prevent the spread of more seed. However, the biology of the weed must be considered before mowing. Some weed species may sprout with increased vigor after being mowed. Also, some species are able to re-sprout from stem or root segments that are left behind after mowing. Brush cutting and weed eating are also mowing techniques that reduce the biomass of the weeds. Repeatedly removing biomass causes reduced vigor in many weed species. This method is usually used in combination with other control methods such as burning or herbicide treatments.

Mulching

Mulch is a layer of material that is spread on the ground. Compared with some other methods of weed control, mulch is relatively simple and inexpensive. Mulching smothers the weeds by excluding light and providing a physical barrier to impede their emergence. Mulching is successful with most annual weeds, however, some perennial weeds are not affected. Mulches may be organic or synthetic. Organic mulches consist of plant by products such as: pine straw, wood chips, green waste, compost, leaves, and grass clippings. Synthetic mulches, also known as ground cover fabric, can be made from materials like polyethylene, polypropylene, or polyester. The effectiveness of mulching is mostly dependent on the material used. Organic and synthetic mulches may be used in combination with each other to increase the amount of weeds controlled.

Tillage

Tillage, also known as cultivation, is the turning over of the soil. This method is more often used in agricultural crops. Tillage can be performed on a small scale with tools such as small, hand pushed rotary tillers or on a large scale with tractor mounted plows. Tillage is able to control weeds because when the soil is overturned, the vegetative parts of the plants are damaged and the root systems are exposed causing desiccation. Generally, the younger the weed is, the more readily it can be controlled with tillage. To control mature perennial weeds, repeated tillage is necessary. By continually destroying new growth and damaging the root system, the weed's food stores are depleted until it can no longer re-sprout. Also, when the soil is overturned, the soil seed bank is disrupted which can cause dormant weed seeds to germinate in the absence of the previous competitors. These new weeds can also be controlled by continued tillage until the soil seed bank is depleted.

Soil Solarization

Soil solarization is a simple method of weed control that is accomplished by covering the soil with a layer of clear or black plastic. The plastic that is covering the ground traps heat energy from the sun and raises the temperature of the soil. Many weed seeds and vegetative propagules are not able to withstand the temperatures and are killed. For this method to be most effective, it should be implemented during the summer months and the soil should be moist. Also, cool season weeds are more susceptible to soil solarization than are warm season weeds. Using black plastic as a cover excludes light which can help to control plants that are growing whereas clear plastic has been shown to produce higher soil temperatures.

Fire

Burning and flaming can be economical and practical methods of weed control if used carefully. For most plants, fire causes the cell walls to rupture when they reach a temperature of 45 °C to 55 °C. Burning is commonly used to control weeds in forests, ditches, and roadsides. Burning can be used to remove accumulated vegetation by destroying the dry, matured plant matter as well as killing the green new growth. Buried weed seeds and plant propagules may also be destroyed during burning, however, dry seeds are much less susceptible to the increased temperature. Flaming is used on a smaller scale and includes the use of a propane torch with a fan tip. Flaming may be used to control weeds along fences and paved areas or places where the soil may be too wet to hoe, dig, or till. Flaming is most effective on young weeds that are less than two inches tall but repeated treatments may control tougher perennial weeds.

Flooding

Flooding is a method of control that requires the area being treated to be saturated at a depth of 15 to 30 cm for a period of 3 to 8 weeks. The saturation of the soil reduces the availability of oxygen to the plant roots thereby killing the weed. This method has been shown to be highly effective in controlling establish perennial weeds and may also suppress annual weeds by reducing the weed seed populations.

Effects of Mechanical Control on the Environment

Mechanical methods of weed control cause physical changes in the immediate environment that may cause positive or negative effects. The suppression of the targeted weeds will open niches in the environment and may also stimulate the growth of other weeds by decreasing their competition and making their environment more favorable. If the niches are not filled by a desirable plant, they will eventually be taken over by another weed. These weed control methods also effect the structure of the soil. The use of mulches can help decrease erosion, decrease water evaporation from the soil, as well as improve the soil structure by increasing the amount of organic matter. Tillage practices can help decrease compaction and aerate the soil. On the other hand, tillage has also been shown to decrease soil moisture, increase soil erosion and runoff, as well as decrease soil microbial populations. Solarization can cause changes in the biological, physical, and chemical properties of the soil. This can cause the soil to be an unfavorable environment for native species which may be beneficial or harmful.

Biological Weed Control

Biological approach to weed control dates back from 1795 when Dactylopius ceylonicus was introduced to control drooping prickly pear (Opuntia vulgaris Miller) over a large area of land; and since then biological control of weeds have been mainly through the classical strategy of introducing natural enemies from areas of co-evolution.

Biological control agents usually target their specific natural enemy weeds. Recently due to certain favorable environmental, health, economic and sustainability reasons; foreign and native organisms that attack weeds are being evaluated for use as biological control agents that may be used to complement conventional methods especially where some weeds have developed resistance to chemical control. Wheeler reported that their international team discovered and tested numerous new species of potential biological control agents that could attack different plant tissues such as defoliators, sap-suckers, stem borers, and leaf- and stem-gall formers. Many successful biological weed control programs in many parts of the world have demonstrated the potency of this approach and support the concept that natural enemies can contribute to the reduction of plant growth and reproduction. Wapshere et al. classified biological approach to weed control as follows: the classical or inoculative method which is based on the introduction of host-specific exotic natural enemies adapted to exotic weeds; the inundative or augmentative method which is based on the mass production and release of native natural enemies usually against native weeds; the conservative method which is based on reducing numbers of native parasites, predators and diseases of native phytophages that feed on native plants; and the broad-spectrum method which is based on the artificial manipulation of the natural enemy population so that the level of attack on the weed is restricted to achieve the de-sired level of control. Also there are three different techniques for applied biocontrol: (1) conservation—protection or maintenance of existing populations of biocontrol agents; (2) augmentation—regular action to increase populations of biocontrol agents, either by periodic releases or by environmental manipulation; and (3) classical biocontrol—the importation and release of exotic biocontrol agents, with the expectation that the agents will become established and further releases will not be necessary.

Louda and Masters stated that despite the positive impact of chemical herbicides in agricultural productivity, complete reliance on chemical control has caused severe problems such as high cost per unit area, decreasing effectiveness, negative impact on plant diversity and increased environmental contamination. He therefore pointed out that the use of biological factors that naturally limit weed populations is one promising alternative. Menaria discussed bioherbicides as an eco-friendly approach to weed man-agement. He explained that the use of chemical herbicides leaves some chemical resi-dues in food commodities which directly or indirectly affect human health. According to him this situation led to the search for alternative methods that are environmentally friendly, and biocontrol has been found a suitable alternative. Green reviewed the po-tential for control using bioherbicides of four important forest weed species in the UK; including bracken, bramble, Japanese knotweed and rhododendron. They concluded that rhododendron is a suitable target weed for control using wood-rotting fungus as a bioherbicide stump treatment; and this is an approach already developed for weedy hardwood species in South Africa, Canada and Netherlands. Clewley et al. analyzed factors associated with control programs (invasive region, native region, plant growth form, target longevity, control agent guild, taxonomy and study duration) in order to identify patterns of control success. They found out that biological control agents sig-nificantly reduced plant size ($28 \pm 4\%$), plant mass ($37 \pm 4\%$), flower and seed produc-tion (35 ± 13 and $42 \pm 9\%$, respectively) and target plant density ($56 \pm 7\%$).

Underlying Principles and Procedures for Biological Weed Control

The underlying principle behind biological approach to weed control is based on some research works that reported that exotic plants become invasive because they have escaped from the insect herbivores and other natural enemies that limit their multiplication and distribution in their native regions; however some other factors may contribute to the tendency for particular plant species to become invasive. Therefore biological control involves using specific natural enemies that can diminish the development and reproduction of their prey organism and put some limitations to them. McFadyen stated that the predominant approach to classical biological weed control involves the importation, colonization, and establishment of exotic natural enemies (predators, parasites, and pathogens) to diminish and maintain exotic pest populations to densities that are economically insignificant.

General Procedures

Some authors have outlined general procedures to be followed when embarking on classical biological weed control programs as follows: (1) evaluate the ecology, economic impact of the weed and potential conflicts of interest; (2) survey the organisms that are already attacking the weed in the new habitat in order to distinguish accidentally introduced agents and so eliminate such from future evaluation; (3) carry out literature search and other forms of survey to identify natural enemies attacking the weed in its native region; (4) screen the possible biological control agents in the foreign country to determine host range and specificity, and to remove nonspecific agents from further consideration; (5) carry out further tests of promising candidates in quarantine after introduction to ensure host specificity and eliminate predators, parasites, and pathogens that may have been introduced with them; (6) embark on mass rearing of host-specific agents; (7) release the host-specific agents; (8) carry out post-release evaluation to determine establishment and effectiveness of agents; and (9) redistribute agents to other areas where control is required.

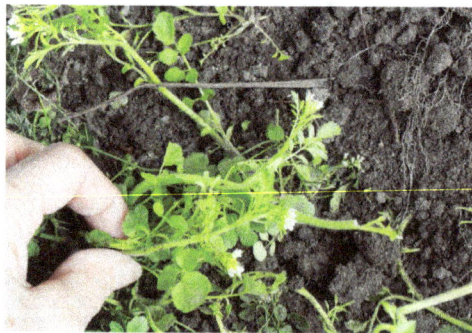

Reasons for Relatively Slow Popularity and Adoption of Biological Weed Control

Weed control practices around the world have shown that the old idea derived from untested opinions; that biological approach to weed control is usually very slow, unpredictable, expensive and mostly unsuccessful is totally not true. Apart from the high initial costs, biological approach to weed control has been known to be relatively cheaper when compared to other methods; however certain factors have slowed down the rate of adoption. These factors include: long time of establishment-usually 20 years or more to ensure success, inadequate or no records of the extent of pre-biological control weed infestations that should serve as a guide for a new biocontrol program, discouraging story of poorly implemented weed bio-control programs. A lot of success stories however have been documented. Lack of information about previous successfully implemented biological control of weeds often lead to untested theories becoming established dogma and this negatively influence the decisions to or not apply it. For instance Mcfadyen stated that it was believed that biological control of trees is difficult, but many examples of trees controlled by insects have been reported. Also classical biological control has been viewed as unsuitable for weeds of annual crops or other frequently disturbed environments, however there are many examples of successful control of crop weeds.

There are evidences showing that some agents introduced for exotic weed control have attacked non target, native plants; and this situation has raised concerns among biological control workers and weed scientists as well as the governments. Opposition to biological approach to control of weeds has also contributed to slowing down the rate of adoption and practice; this is because weed control scientists believe that it is difficult to estimate the cost or the feasibility of biocontrol. Based on a study carried out in South Africa, it was reported that some of the weed biocontrol projects have provided practical solutions to problems e.g. the development of Stumpout for the treatment of wattle stumps and the use of *C. gloeosporioides* for the control of *H. sericea*. However other projects have been less successful and have resulted in the rejection of potential agents for various reasons and these include *C. albofundus on A. mearnsii, X. campestris on M. aquaticum and G. nitens on R. cuneifolius*. Vurro and Evans identified legislative hurdles, technological and commercial constraints as limitations to the adoption of biological weed control in Europe. Olckers stated that limited budgets in many countries have also helped to slow than the rate of adoption and practice of biological approach to weed control.

Examples of Successful Biological Control of Weeds with Introduced Insects and Pathogens

One thousand one hundred and forty-four individuals (mostly entomologists and plant pathologists) have ever attended the International Symposia on Biological Control of Weeds (ISBCWs); and out of these, 450–550 weed biological control experts have been actively involved in research and development efforts over the last 50 years mainly from USA, Canada, Australia, South Africa and New Zealand. McFadyen reported that biological approach to weed control has a long history and a good success rate of 94. A comprehensive list of agents and their target weeds have been documented by Winston et al. Culliney presented potential benefits estimated for some proposed or initiated biological control programs targeting invasive weeds. Frequently cited examples of successful approach to biological weed control are the prickly pear cacti (*Opuntia; spp.)* in Australia, eradicated by an imported moth (*Cactoblastis cactorum*) and rangeland in California, Oregon, Washington, and British Columbia controlled by St. John's wort *Hypericum perforatum* (millepertuis perforé). Mcfadyen presented a list of 41 weds which have successfully been controlled using introduced insects and pathogens and another three weeds also controlled by introduced fungi applied as mycoherbicides. He further stated that many of these successes have been repeated in other countries and continents. Julien presented a list of both successful and failed cases of biological weed control; this included the introduction of 225 organisms against 111 weed species, and 178 insects and 6 mites. Palmer et al. reported that 43 new arthropod or pathogen agents were released in 19 projects; and that effective biological control was achieved in several projects with the outstanding successes being the control of rubber vine, *Cryptostegia grandiflora*, and bridal creeper, *Asparagus asparagoides*.

When is Weed Biological Control Successful?

Information collated on weed impacts before the initiation of a biological control program is necessary to provide baseline data and devise performance criteria with which the program can subsequently be evaluated. For avoidance of confusion on when a biological control could be viewed as successful or not, Hoffmann stated that an implementation of a particular biological control will be termed successful when: complete-when no other control method is required or used, at least in areas where the agentis established; substantial-where other methods are needed but the effort required is reduced (e.g. less herbicide or less frequent application); and negligible-where despite damage inflicted by agents, control of the weed is still dependent on other control measures. Complete control does not imply total eradication of the weed; rather it means that control measures are not required anymore specifically against the target weed, and that crop or pasture yield losses will not be attributed mainly to this weed. Substantial control involves situations where control may be complete in some seasons and over part of the weed's range, as well as cases where the control achieved is widespread and economically significant but the weed is still a major problem. It is therefore concluded that successful implementation of biological approach to weed control is the successful control of the weed, and not necessarily the successful establishment of individual agents released against the weed. Successful biological control depends on three factors: the extent to which each individual agent can limit the targeted plant; the ecology of the agent as it affects its ability to populate and spread easily in the new environment; and the ecology of the weed, which determines if the total damage that can be caused by the agent can significantly reduce its population. Because agents always need some surviving predator plants to complete their life cycle, biological control will not usually totally eradicate their target weeds. In essence a successful biological control program reduces the potency and population of the target weed and usually in conjunction with other control methods as part of an overall integrated weed management scheme which is recommended.

Things to Consider When Making the Choice of Agents to be Introduced to Control Weeds

Gassmann reported that selection of potential agents in the last decades has been mainly based on the population biology of the weed, impact studies of agents on the plant and the combined effect of herbivory and plant competition. Palmer et al. stated that agent selection is highly dependent on the type of weed, its reproductive system, on the ecological, abiotic and management context in which that weed occurs, and on the acceptable goals and impact thresholds required of a biological control program. Generally, factors to be considered in selecting agents include the following: the agent must target a particular plant species, must have high level of predation and parasitism on the host plant and its entire population, must be prolific, must be able to thrive in all habitats and climates where the weed exists and should be able to spread easily and widely, must be a strong colonizer, the overall cost of introducing the agent must be cheaper compared to other control methods, the technology that will be involved in introducing and managing the agent must be as simple as possible, must as much as possible maintain natural biodiversity, sufficient number of individuals must be released, plant phenology (effect of periodic plant life cycle events) must be favorable. To be considered a good candidate for biological control, a weed should be non-native, present in numbers and densities greater than in its native range and numerous enough to cause environmental or economic damage, the weed should also be present over a broad geographic range, have few or no redeeming or beneficial qualities, have taxonomic characteristics sufficiently distinct from those of economically important and native plant. Furthermore, the weed should occur in relatively undisturbed areas to allow for the establishment of biological control agents, cultivation, mowing and other disturbances can have a destructive effect on many arthropod biocontrol agents. Inundative biocontrol agents such as bacteria and fungi are less sensitive to these types of disturbances so may be used in cropland.

Steps to Identifying and Introducing Biological Control Agents

The study of insect attributes and fitness traits, the influence of plant resources on insect performance, and the construction of comparative life-tables, are the first steps towards an improvement of the success rate of biological weed control. Generally, steps to identifying and introducing biological control agents include: (1) identify target weeds; (2) identify control agents and determine the level of specialization; (3) apply controlled release of the agents; (4) apply full release and determine optimal release sites; (5) for the case of classical methods, monitor release sites; (6) apply redistribution for the case of classical methods (7) and maintain control agent populations.

Chemical Weed Control

Chemical weed control refers to any technique that involves the application of a chemical (herbicide) to weeds or soil to control the germination or growth of the weed species. In economic terms, chemical control of weeds is a very large industry and there are scores of examples of chemical weed control products. Common examples of chemicals used to control weeds in forages are 2,4-DB; EPTC; bromoxynil; paraquat. Knowledge of weed seed characteristics, morphology, ontogeny, nature of competition and degree of association with crops are pre-requisite for suggesting some efficient weed control measures. It makes the users/scientists quite acquainted with the nature and spectrum of weeds existing in the crop fields and accordingly guides them to adopt certain measures. Identification and naming of a particular weed based on its genus, species or certain biological characters may not be much useful to users since weed control usually, unless specific weed problem in certain area, aims at composite weed culture and not on individual species of weeds. Therefore, some common characteristics of the species, which are clearly visible and easily understandable by users, are to be exploited for making of their classes/groups and for recommending suitable control measures.

Chemicals that are used to kill plants or weeds are called herbicides. A proper technical know-how is a pre-requisite for successful adoption of chemical method of weed control so-called herbicide technology. One has to exercise a lot of caution while using the herbicide for uniform application as well as higher herbicide efficiency. Herbicide selectivity and its dose, time and method of application are of paramount importance/ consideration before applying to a crop.

Herbicides

Herbicides also commonly known as weedkillers, are substances used to control unwanted plants. Selective herbicides control specific weed species, while leaving the desired crop relatively unharmed, while non-selective herbicides (sometimes called total weedkillers in commercial products) can be used to clear waste ground, industrial and construction sites, railways and railway embankments as they kill all plant material with which they come into contact. Apart from selective/non-selective, other important distinctions include persistence (also known as residual action: how long the product stays in place and remains active), means of uptake (whether it is absorbed by above-ground foliage only, through the roots, or by other means), and mechanism of action (how it works). Historically, products such as common salt and other metal salts were used as herbicides, however these have gradually fallen out of favor and in some countries a number of these are banned due to their persistence in soil, and toxicity and groundwater contamination concerns. Herbicides have also been used in warfare and conflict.

Modern herbicides are often synthetic mimics of natural plant hormones which interfere with growth of the target plants. The term organic herbicide has come to mean herbicides intended for organic farming. Some plants also produce their own natural herbicides, such as the genus *Juglans* (walnuts), or the tree of heaven; such action of natural herbicides, and other related chemical interactions, is called allelopathy. Due to herbicide resistance - a major concern in agriculture - a number of products combine herbicides with different means of action. Integrated pest management may use herbicides alongside other pest control methods.

In the US in 2007, about 83% of all herbicide usage, determined by weight applied, was in agriculture. In 2007, world pesticide expenditures totaled about $39.4 billion; herbicides were about 40% of those sales and constituted the biggest portion, followed by insecticides, fungicides, and other types. Herbicide is also used in forestry, where certain formulations have been found to suppress hardwood varieties in favour of conifers after a clearcut, as well as pasture systems, and management of areas set aside as wildlife habitat.

First Herbicides

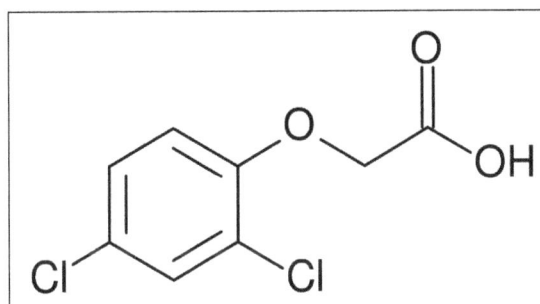

2,4-D, the first chemical herbicide, was discovered during the Second World War.

Although research into herbicides began in the early 20th century, the first major breakthrough was the result of research conducted in both the UK and the US during the Second World War into the potential use of herbicides in war. The first modern herbicide, 2,4-D, was first discovered and synthesized by W. G. Templeman at Imperial Chemical Industries. In 1940, he showed that "Growth substances applied appropriately would kill certain broad-leaved weeds in cereals without harming the crops." By 1941, his team succeeded in synthesizing the chemical. In the same year, Pokorny in the US achieved this as well.

Independently, a team under Juda Hirsch Quastel, working at the Rothamsted Experimental Station made the same discovery. Quastel was tasked by the Agricultural Research Council (ARC) to discover methods for improving crop yield. By analyzing soil as a dynamic system, rather than an inert substance, he was able to apply techniques such as perfusion. Quastel was able to quantify the influence of various plant hormones, inhibitors and other chemicals on the activity of microorganisms in the soil and assess their direct impact on plant growth. While the full work of the unit remained secret, certain discoveries were developed for commercial use after the war, including the 2,4-D compound.

When 2,4-D was commercially released in 1946, it triggered a worldwide revolution in agricultural output and became the first successful selective herbicide. It allowed for greatly enhanced weed control in wheat, maize (corn), rice, and similar cereal grass crops, because it kills dicots (broadleaf plants), but not most monocots (grasses). The low cost of 2,4-D has led to continued usage today, and it remains one of the most commonly used herbicides in the world. Like other acid herbicides, current formulations use either an amine salt (often trimethylamine) or one of many esters of the parent compound. These are easier to handle than the acid.

Further Discoveries

The triazine family of herbicides, which includes atrazine, were introduced in the 1950s; they have the current distinction of being the herbicide family of greatest concern regarding groundwater contamination. Atrazine does not break down readily (within a few weeks) after being applied to soils of above neutral pH. Under alkaline soil conditions, atrazine may be carried into the soil profile as far as the water table by soil water following rainfall causing the aforementioned contamination. Atrazine is thus said to have "carryover", a generally undesirable property for herbicides.

Glyphosate (Roundup) was introduced in 1974 for nonselective weed control. Following the development of glyphosate-resistant crop plants, it is now used very extensively for selective weed control in growing crops. The pairing of the herbicide with the resistant seed contributed to the consolidation of the seed and chemistry industry in the late 1990s.

Many modern herbicides used in agriculture and gardening are specifically formulated to decompose within a short period after application. This is desirable, as it allows crops and plants to be planted afterwards, which could otherwise be affected by the herbicide. However, herbicides with low residual activity (i.e., that decompose quickly) often do not provide season-long weed control and do not ensure that weed roots are killed beneath construction and paving (and cannot emerge destructively in years to come), therefore there remains a role for weedkiller with high levels of persistence in the soil.

Terminology

Herbicides are classified/grouped in various ways e.g. according to the activity, timing of application, method of application, mechanism of action, chemical family. This gives rise to a considerable level of terminology related to herbicides and their use.

Intended Outcome

- Control is the destruction of unwanted weeds, or the damage of them to the point where they are no longer competitive with the crop.

- Suppression is incomplete control still providing some economic benefit, such as reduced competition with the crop.

- Crop safety, for selective herbicides, is the relative absence of damage or stress to the crop. Most selective herbicides cause some visible stress to crop plants.

- Defoliant, similar to herbicides, but designed to remove foliage (leaves) rather than kill the plant.

Selectivity

- Selective herbicides control or suppress certain plants without affecting the growth of other plants species. Selectivity may be due to translocation, differential absorption, physical (morphological) or physiological differences between plant species. 2,4-D, mecoprop, dicamba control many broadleaf weeds but remain ineffective against turfgrasses.

- Non-selective herbicides are not specific in acting against certain plant species and control all plant material with which they come into contact. They are used to clear industrial sites, waste ground, railways and railway embankments. Paraquat, glufosinate, glyphosate are non-selective herbicides.

Timing of Application

- Preplant: Preplant herbicides are nonselective herbicides applied to soil before planting. Some preplant herbicides may be mechanically incorporated into the soil. The objective for incorporation is to prevent dissipation through photodecomposition and volatility. The herbicides kill weeds as they grow through the herbicide treated zone. Volatile herbicides have to be incorporated into the soil before planting the pasture. Agricultural crops grown in soil treated with a preplant herbicide include tomatoes, corn, soybeans and strawberries. Soil fumigants like metam-sodium and dazomet are in use as preplant herbicides.

- Preemergence: Preemergence herbicides are applied before the weed seedlings emerge through the soil surface. Herbicides do not prevent weeds from germinating but they kill weeds as they grow through the herbicide treated zone by affecting the cell division in the emerging seedling. Dithopyr and pendimethalin are preemergence herbicides. Weeds that have already emerged before application or activation are not affected by pre-herbicides as their primary growing point escapes the treatment.

- Postemergence: These herbicides are applied after weed seedlings have emerged through the soil surface. They can be foliar or root absorbed, selective or non-selective, contact or systemic. Application of these herbicides is avoided during rain because the problem of being washed off to the soil makes it ineffective. 2,4-D is a selective, systemic, foliar absorbed postemergence herbicide.

Method of Application

- Soil applied: Herbicides applied to the soil are usually taken up by the root or shoot of the emerging seedlings and are used as preplant or preemergence treatment. Several factors influence the effectiveness of soil-applied herbicides. Weeds absorb herbicides by both passive and active mechanism. Herbicide adsorption to soil colloids or organic matter often reduces its amount available for weed absorption. Positioning of herbicide in correct layer of soil is very important, which can be achieved mechanically and by rainfall. Herbicides on the soil surface are subjected to several processes that reduce their availability. Volatility and photolysis are two common processes that reduce the availability of herbicides. Many soil applied herbicides are absorbed through plant shoots while they are still underground leading to their death or injury. EPTC and trifluralin are soil applied herbicides.

- Foliar applied: These are applied to portion of the plant above the ground and are absorbed by exposed tissues. These are generally postemergence herbicides and can either be translocated (systemic) throughout the plant or remain at specific site (contact). External barriers of plants like cuticle, waxes, cell wall etc. affect herbicide absorption and action. Glyphosate, 2,4-D and dicamba are foliar applied herbicide.

Persistence

- Residual activity: An herbicide is described as having low residual activity if it is neutralized within a short time of application (within a few weeks or months) typically this is due to rainfall, or by reactions in the soil. An herbicide described as having high residual activity will remain potent for a long term in the soil. For some compounds, the residual activity can leave the ground almost permanently barren.

Mechanism of Action

Herbicides are often classified according to their site of action, because as a general rule, herbicides within the same site of action class will produce similar symptoms on susceptible plants. Classification based on site of action of herbicide is comparatively better as herbicide resistance management can be handled more properly and effectively. Classification by mechanism of action (MOA) indicates the first enzyme, protein, or biochemical step affected in the plant following application.

List of mechanisms found in modern herbicides:

- ACCase inhibitors: Acetyl coenzyme A carboxylase (ACCase) is part of the first step of lipid synthesis. Thus, ACCase inhibitors affect cell membrane production in the meristems of the grass plant. The ACCases of grasses are sensitive to these herbicides, whereas the ACCases of dicot plants are not.

- ALS inhibitors: the acetolactate synthase (ALS) enzyme (also known as acetohydroxyacid synthase, or AHAS) is the first step in the synthesis of the branched-chain amino acids (valine, leucine, and isoleucine). These herbicides slowly starve affected plants of these amino acids, which eventually leads to inhibition of DNA synthesis. They affect grasses and dicots alike. The ALS inhibitor family includes various sulfonylureas (SUs) (such as Flazasulfuron and Metsulfuron-methyl), imidazolinones (IMIs), triazolopyrimidines (TPs), pyrimidinyl oxybenzoates (POBs), and sulfonylamino carbonyl triazolinones (SCTs). The ALS biological pathway exists only in plants and not animals, thus making the ALS-inhibitors among the safest herbicides.

- EPSPS inhibitors: Enolpyruvylshikimate 3-phosphate synthase enzyme (EPSPS) is used in the synthesis of the amino acids tryptophan, phenylalanine and tyrosine. They affect grasses and dicots alike. Glyphosate (Roundup) is a systemic EPSPS inhibitor inactivated by soil contact.

- The discovery of synthetic auxins inaugurated the era of organic herbicides. They were discovered in the 1940s after long study of the plant growth regulator auxin. Synthetic auxins mimic in some way this plant hormone. They have several points of action on the cell membrane, and are effective in the control of dicot plants. 2,4-D and 2,4,5-T are synthetic auxin herbicides.

- Photosystem II inhibitors reduce electron flow from water to $NADP^+$ at the pho-tochemical step in photosynthesis. They bind to the Qb site on the D1 protein, and prevent quinone from binding to this site. Therefore, this group of compounds causes electrons to accumulate on chlorophyll molecules. As a consequence, oxidation reactions in excess of those normally tolerated by the cell occur, and the plant dies. The triazine herbicides (including atrazine) and urea derivatives (diuron) are photosystem II inhibitors.

- Photosystem I inhibitors steal electrons from the normal pathway through FeS to Fdx to NADP⁺ leading to direct discharge of electrons on oxygen. As a result, reactive oxygen species are produced and oxidation reactions in excess of those normally tolerated by the cell occur, leading to plant death. Bipyridinium herbicides (such as diquat and paraquat) inhibit the FeS to Fdx step of that chain, while diphenyl ether herbicides (such as nitrofen, nitrofluorfen, and acifluorfen) inhibit the Fdx to NADP⁺ step.

- HPPD inhibitors inhibit 4-Hydroxyphenylpyruvate dioxygenase, which are involved in tyrosine breakdown. Tyrosine breakdown products are used by plants to make carotenoids, which protect chlorophyll in plants from being destroyed by sunlight. If this happens, the plants turn white due to complete loss of chlorophyll, and the plants die. Mesotrione and sulcotrione are herbicides in this class; a drug, nitisinone, was discovered in the course of developing this class of herbicides.

Herbicide Group

One of the most important methods for preventing, delaying, or managing resistance is to reduce the reliance on a single herbicide mode of action. To do this, farmers must know the mode of action for the herbicides they intend to use, but the relatively complex nature of plant biochemistry makes this difficult to determine. Attempts were made to simplify the understanding of herbicide mode of action by developing a classification system that grouped herbicides by mode of action. Eventually the Herbicide Resistance Action Committee (HRAC) and the Weed Science Society of America (WSSA) developed a classification system. The WSSA and HRAC systems differ in the group designation. Groups in the WSSA and the HRAC systems are designated by numbers and letters, respectively. The goal for adding the "Group" classification and mode of action to the herbicide product label is to provide a simple and practical approach to deliver the information to users. This information will make it easier to develop educational material that is consistent and effective. It should increase user's awareness of herbicide mode of action and provide more accurate recommendations for resistance management. Another goal is to make it easier for users to keep records on which herbicide mode of actions are being used on a particular field from year to year.

Chemical Family

Detailed investigations on chemical structure of the active ingredients of the registered herbicides showed that some moieties (moiety is a part of a molecule that may include either whole functional groups or parts of functional groups as substructures; a functional group has similar chemical properties whenever it occurs in different compounds) have the same mechanisms of action. According to Forouzesh *et al.* 2015, these moieties have been assigned to the names of chemical families and active ingredients are then classified within the chemical families accordingly. Knowing about herbicide chemical family grouping could serve as a short-term strategy for managing resistance to site of action.

Use and Application

Herbicides being sprayed from the spray arms of a tractor in North Dakota.

Most herbicides are applied as water-based sprays using ground equipment. Ground equipment varies in design, but large areas can be sprayed using self-propelled sprayers equipped with long booms, of 60 to 120 feet (18 to 37 m) with spray nozzles spaced every 20–30 inches (510–760 mm) apart. Towed, handheld, and even horse-drawn sprayers are also used. On large areas, herbicides may also at times be applied aerially using helicopters or airplanes, or through irrigation systems (known as chemigation).

A further method of herbicide application developed around 2010, involves ridding the soil of its active weed seed bank rather than just killing the weed. This can successfully treat annual plants but not perennials. Researchers at the Agricultural Research Service found that the application of herbicides to fields late in the weeds' growing season greatly reduces their seed production, and therefore fewer weeds will return the following season. Because most weeds are annuals, their seeds will only survive in soil for a year or two, so this method will be able to destroy such weeds after a few years of herbicide application.

Weed-wiping may also be used, where a wick wetted with herbicide is suspended from a boom and dragged or rolled across the tops of the taller weed plants. This allows treatment of taller grassland weeds by direct contact without affecting related but desirable shorter plants in the grassland sward beneath. The method has the benefit of avoiding spray drift. In Wales, a scheme offering free weed-wiper hire was launched in 2015 in an effort to reduce the levels of MCPA in water courses.

Misuse and Misapplication

Herbicide volatilisation or spray drift may result in herbicide affecting neighboring fields or plants, particularly in windy conditions. Sometimes, the wrong field or plants may be sprayed due to error.

Use Politically, Militarily and in Conflict

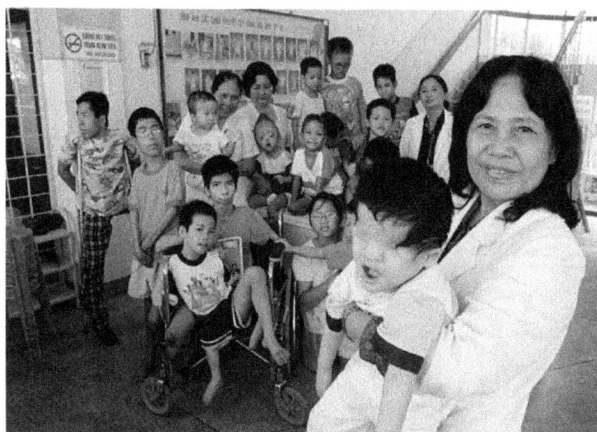

Handicapped children in Vietnam, most of them victims of Agent Orange.

Although herbicidal warfare use chemical substances, its main purpose is to disrupt agricultural food production and to destroy plants which provide cover or concealment to the enemy.

The use of herbicides as a chemical weapon by the U.S. military during the Vietnam War has left tangible, long-term impacts upon the Vietnamese people that live in Vietnam. For instance, it led to 3 million Vietnamese people suffering health problems, one million birth defects caused directly by exposure to Agent Orange, and 24% of the area of Vietnam being defoliated.

Health and Environmental Effects

Herbicides have widely variable toxicity in addition to acute toxicity arising from ingestion of a significant quantity rapidly, and chronic toxicity arising from environmental and occupational exposure over long periods. Much public suspicion of herbicides revolves around a confusion between valid statements of *acute* toxicity as opposed to equally valid statements of lack of *chronic* toxicity at the recommended levels of usage. For instance, while glyphosate formulations with tallowamine *adjuvants* are acutely toxic, their use was found to be uncorrelated with any health issues like cancer in a massive US Department of Health study on 90,000 members of farmer families for over a period of 23 years. That is, the study shows lack of chronic toxicity, but cannot question the herbicide's acute toxicity.

Some herbicides cause a range of health effects ranging from skin rashes to death. The pathway of attack can arise from intentional or unintentional direct consumption, improper application resulting in the herbicide coming into direct contact with people or wildlife, inhalation of aerial sprays, or food consumption prior to the labelled preharvest interval. Under some conditions, certain herbicides can be transported via leaching or surface runoff to contaminate groundwater or distant surface water sources. Generally, the conditions that promote herbicide transport include intense storm events (particularly shortly after application) and soils with limited capacity to adsorb or retain the herbicides. Herbicide properties that increase likelihood of transport include persistence (resistance to degradation) and high water solubility.

Phenoxy herbicides are often contaminated with dioxins such as TCDD; research has suggested such contamination results in a small rise in cancer risk after occupational exposure to these herbicides. Triazine exposure has been implicated in a likely relationship to increased risk of breast cancer, although a causal relationship remains unclear.

Herbicide manufacturers have at times made false or misleading claims about the safety of their products. Chemical manufacturer Monsanto Company agreed to change its advertising after pressure from New York attorney general Dennis Vacco; Vacco complained about misleading claims that its spray-on glyphosate-based herbicides, including Roundup, were safer than table salt and "practically non-toxic" to mammals, birds, and fish (though proof that this was ever said is hard to find). Roundup is toxic and has resulted in death after being ingested in quantities ranging from 85 to 200 ml, although it has also been ingested in quantities as large as 500 ml with only mild or moderate symptoms. The manufacturer of Tordon 101 (Dow AgroSciences, owned by the Dow Chemical Company) has claimed Tordon 101 has no effects on animals and insects, in spite of evidence of strong carcinogenic activity of the active ingredient Picloram in studies on rats.

The risk of Parkinson's disease has been shown to increase with occupational exposure to herbicides and pesticides. The herbicide paraquat is suspected to be one such factor.

All commercially sold, organic and nonorganic herbicides must be extensively tested prior to approval for sale and labeling by the Environmental Protection Agency. However, because of the large number of herbicides in use, concern regarding health effects is significant. In addition to health effects caused by herbicides themselves, commercial herbicide mixtures often contain other chemicals, including inactive ingredients, which have negative impacts on human health.

Ecological Effects

Commercial herbicide use generally has negative impacts on bird populations, although the impacts are highly variable and often require field studies to predict accurately. Laboratory studies have at times overestimated negative impacts on birds due to toxicity, predicting serious problems that were not observed in the field. Most observed effects are due not to toxicity, but to habitat changes and the decreases in abundance of species on which birds rely for food or shelter. Herbicide use in silviculture, used to favor certain types of growth following clearcutting, can cause significant drops in bird populations. Even when herbicides which have low toxicity to birds are used, they decrease the abundance of many types of vegetation on which the birds rely. Herbicide use in agriculture in Britain has been linked to a decline in seed-eating bird species which rely on the weeds killed by the herbicides. Heavy use of herbicides in neotropical agricultural areas has been one of many factors implicated in limiting the usefulness of such agricultural land for wintering migratory birds.

Frog populations may be affected negatively by the use of herbicides as well. While some studies have shown that atrazine may be a teratogen, causing demasculinization in male frogs, the U.S. Environmental Protection Agency (EPA) and its independent Scientific Advisory Panel (SAP) examined all available studies on this topic and concluded that "atrazine does not adversely affect amphibian gonadal development based on a review of laboratory and field studies".

Scientific Uncertainty of Full Extent of Herbicide Effects

The health and environmental effects of many herbicides is unknown, and even the scientific community often disagrees on the risk. For example, a 1995 panel of 13 scientists reviewing studies on the carcinogenicity of 2,4-D had divided opinions on the likelihood 2,4-D causes cancer in humans. As of 1992, studies on phenoxy herbicides were too few to accurately assess the risk of many types of cancer from these herbicides, even though evidence was stronger that exposure to these herbicides is associated with increased risk of soft tissue sarcoma and non-Hodgkin lymphoma. Furthermore, there is some suggestion that herbicides can play a role in sex reversal of certain organisms that experience temperature-dependent sex determination, which could theoretically alter sex ratios.

Resistance

Weed resistance to herbicides has become a major concern in crop production worldwide. Resistance to herbicides is often attributed to lack of rotational programmes of herbicides and to continuous applications of herbicides with the same sites of action. Thus, a true understanding of the sites of action of herbicides is essential for strategic planning of herbicide-based weed control.

Plants have developed resistance to atrazine and to ALS-inhibitors, and more recently, to glyphosate herbicides. Marestail is one weed that has developed glyphosate resistance. Glyphosate-resistant weeds are present in the vast majority of soybean, cotton and corn farms in some U.S. states. Weeds that can resist multiple other herbicides are spreading. Few new herbicides are near commercialization, and none with a molecular mode of action for which there is no resistance. Because most herbicides could not kill all weeds, farmers rotated crops and herbicides to stop resistant weeds. During its initial years, glyphosate was not subject to resistance and allowed farmers to reduce the use of rotation.

A family of weeds that includes waterhemp (Amaranthus rudis) is the largest concern. A 2008-9 survey of 144 populations of waterhemp in 41 Missouri counties revealed glyphosate resistance in 69%. Weeds from some 500 sites throughout Iowa in 2011 and 2012 revealed glyphosate resistance in approximately 64% of waterhemp samples. The use of other killers to target "residual" weeds has become common, and may be sufficient to have stopped the spread of resistance From 2005 through 2010 researchers discovered 13 different weed species that had developed resistance to glyphosate. But since then only two more have been discovered. Weeds resistant to multiple herbicides with completely different biological action modes are on the rise. In Missouri, 43% of samples were resistant to two different herbicides; 6% resisted three; and 0.5% resisted four. In Iowa 89% of waterhemp samples resist two or more herbicides, 25% resist three, and 10% resist five.

For southern cotton, herbicide costs has climbed from between $50 and $75 per hectare a few years ago to about $370 per hectare in 2013. Resistance is contributing to a massive shift away from growing cotton; over the past few years, the area planted with cotton has declined by 70% in Arkansas and by 60% in Tennessee. For soybeans in Illinois, costs have risen from about $25 to $160 per hectare.

Dow, Bayer CropScience, Syngenta and Monsanto are all developing seed varieties resistant to herbicides other than glyphosate, which will make it easier for farmers to use alternative weed killers. Even though weeds have already evolved some resistance to those herbicides, Powles says the new seed-and-herbicide combos should work well if used with proper rotation.

Biochemistry of Resistance

Resistance to herbicides can be based on one of the following biochemical mechanisms:

- Target-site resistance: This is due to a reduced (or even lost) ability of the herbicide to bind to its target protein. The effect usually relates to an enzyme with a crucial function in a metabolic pathway, or to a component of an electron-transport system. Target-site resistance may also be caused by an overexpression of the target enzyme (via gene amplification or changes in a gene promoter).

- Non-target-site resistance: This is caused by mechanisms that reduce the amount of herbicidal active compound reaching the target site. One important mechanism is an enhanced metabolic detoxification of the herbicide in the weed, which leads to insufficient amounts of the active substance reaching the target site. A reduced uptake and translocation, or sequestration of the herbicide, may also result in an insufficient herbicide transport to the target site.

- Cross-resistance: In this case, a single resistance mechanism causes resistance to several herbicides. The term target-site cross-resistance is used when the herbicides bind to the same target site, whereas non-target-site cross-resistance is due to a single non-target-site mechanism (e.g., enhanced metabolic detoxification) that entails resistance across herbicides with different sites of action.

- Multiple resistance: In this situation, two or more resistance mechanisms are present within individual plants, or within a plant population.

Resistance Management

Worldwide experience has been that farmers tend to do little to prevent herbicide resistance developing, and only take action when it is a problem on their own farm or neighbor's. Careful observation is important so that any reduction in herbicide efficacy can be detected. This may indicate evolving resistance. It is vital that resistance is detected at an early stage as if it becomes an acute, whole-farm problem, options are more limited and greater expense is almost inevitable. Table lists factors which enable the risk of resistance to be assessed. An essential pre-requisite for confirmation of resistance is a good diagnostic test. Ideally this should be rapid, accurate, cheap and accessible. Many diagnostic tests have been developed, including glasshouse pot assays, petri dish assays and chlorophyll fluorescence. A key component of such tests is that the response of the suspect population to a herbicide can be compared with that of known susceptible and resistant standards under controlled conditions. Most cases of herbicide resistance are a consequence of the repeated use of herbicides, often in association with crop monoculture and reduced cultivation practices. It is necessary, therefore, to modify these practices in order to prevent or delay the onset of resistance or to control existing resistant populations. A key objective should be the reduction in selection pressure. An integrated weed management (IWM) approach is required, in which as many tactics as possible are used to combat weeds. In this way, less reliance is placed on herbicides and so selection pressure should be reduced.

Optimising herbicide input to the economic threshold level should avoid the unnecessary use of herbicides and reduce selection pressure. Herbicides should be used to their greatest potential by ensuring that the timing, dose, application method, soil and climatic conditions are optimal for good activity. In the UK, partially resistant grass weeds such as *Alopecurus myosuroides* (blackgrass) and *Avena* spp. (wild oat) can often be controlled adequately when herbicides are applied at the 2-3 leaf stage, whereas later applications at the 2-3 tiller stage can fail badly. Patch spraying, or applying herbicide to only the badly infested areas of fields, is another means of reducing total herbicide use.

Table: Agronomic factors influencing the risk of herbicide resistance development.

Factor	Low risk	High risk
Cropping system	Good rotation	Crop monoculture
Cultivation system	Annual ploughing	Continuous minimum tillage
Weed control	Cultural only	Herbicide only
Herbicide use	Many modes of action	Single modes of action
Control in previous years	Excellent	Poor
Weed infestation	Low	High
Resistance in vicinity	Unknown	Common

Approaches to Treating Resistant Weeds

Alternative Herbicides

When resistance is first suspected or confirmed, the efficacy of alternatives is likely to be the first consideration. The use of alternative herbicides which remain effective on resistant populations can be a successful strategy, at least in the short term. The effectiveness of alternative herbicides will be highly dependent on the extent of cross-resistance. If there is resistance to a single group of herbicides, then the use of herbicides from other groups may provide a simple and effective solution, at least in the short term. For example, many triazine-resistant weeds have been readily controlled by the use of alternative herbicides such as dicamba or glyphosate. If resistance extends to more than one herbicide group, then choices are more limited. It should not be assumed that resistance will automatically extend to all herbicides with the same mode of action, although it is wise to assume this until proved otherwise. In many weeds the degree of cross-resistance between the five groups of ALS inhibitors varies considerably. Much will depend on the resistance mechanisms present, and it should not be assumed that these will necessarily be the same in different populations of the same species. These differences are due, at least in part, to the existence of different mutations conferring target site resistance. Consequently, selection for different mutations may result in different patterns of cross-resistance. Enhanced metabolism can affect even closely related herbicides to differing degrees. For example, populations of *Alopecurus myosuroides* (blackgrass) with an enhanced metabolism mechanism show resistance to pendimethalin but not to trifluralin, despite both being dinitroanilines. This is due to differences in the vulnerability of these two herbicides to oxidative metabolism. Consequently, care is needed when trying to predict the efficacy of alternative herbicides.

Mixtures and Sequences

The use of two or more herbicides which have differing modes of action can reduce the selection for resistant genotypes. Ideally, each component in a mixture should:

- Be active at different target sites.

- Have a high level of efficacy.

- Be detoxified by different biochemical pathways.

- Have similar persistence in the soil (if it is a residual herbicide).

- Exert negative cross-resistance.

- Synergise the activity of the other component.

No mixture is likely to have all these attributes, but the first two listed are the most important. There is a risk that mixtures will select for resistance to both components in the longer term. One practical advantage of sequences of two herbicides compared with mixtures is that a better appraisal of the efficacy of each herbicide component is possible, provided that sufficient time elapses between each application. A disadvantage with sequences is that two separate applications have to be made and it is possible that the later application will be less effective on weeds surviving the first application. If these are resistant, then the second herbicide in the sequence may increase selection for resistant individuals by killing the susceptible plants which were damaged but not killed by the first application, but allowing the larger, less affected, resistant plants to survive. This has been cited as one reason why ALS-resistant *Stellaria media* has evolved in Scotland recently (2000), despite the regular use of a sequence incorporating mecoprop, a herbicide with a different mode of action.

Herbicide Rotations

Rotation of herbicides from different chemical groups in successive years should reduce selection for resistance. This is a key element in most resistance prevention programmes. The value of this approach depends on the extent of cross-resistance, and whether multiple resistance occurs owing to the presence of several different resistance mechanisms. A practical problem can be the lack of awareness by farmers of the different groups of herbicides that exist. In Australia a scheme has been introduced in which identifying letters are included on the product label as a means of enabling farmers to distinguish products with different modes of action.

Farming Practices and Resistance: A Case Study

Herbicide resistance became a critical problem in Australian agriculture, after many Australian sheep farmers began to exclusively grow wheat in their pastures in the 1970s.

Introduced varieties of ryegrass, while good for grazing sheep, compete intensely with wheat. Ryegrasses produce so many seeds that, if left unchecked, they can completely choke a field. Herbicides provided excellent control, while reducing soil disrupting because of less need to plough. Within little more than a decade, ryegrass and other weeds began to develop resistance. In response Australian farmers changed methods. By 1983, patches of ryegrass had become immune to Hoegrass, a family of herbicides that inhibit an enzyme called acetyl coenzyme A carboxylase.

Ryegrass populations were large, and had substantial genetic diversity, because farmers had planted many varieties. Ryegrass is cross-pollinated by wind, so genes shuffle frequently. To control its distribution farmers sprayed inexpensive Hoegrass, creating selection pressure. In addition, farmers sometimes diluted the herbicide in order to save money, which allowed some plants to survive application. When resistance appeared farmers turned to a group of herbicides that block acetolactate synthase. Once again, ryegrass in Australia evolved a kind of "cross-resistance" that allowed it to rapidly break down a variety of herbicides. Four classes of herbicides become ineffective within a few years. In 2013 only two herbicide classes, called Photosystem II and long-chain fatty acid inhibitors, were effective against ryegrass.

List of Common Herbicides

Synthetic Herbicides

- 2,4-D is a broadleaf herbicide in the phenoxy group used in turf and no-till field crop production. Now, it is mainly used in a blend with other herbicides to allow lower rates of herbicides to be used; it is the most widely used herbicide in the world, and third most commonly used in the United States. It is an example of synthetic auxin (plant hormone).

- Aminopyralid is a broadleaf herbicide in the pyridine group, used to control weeds on grassland, such as docks, thistles and nettles. It is notorious for its ability to persist in compost.

- Atrazin, a triazine herbicide, is used in corn and sorghum for control of broadleaf weeds and grasses. Still used because of its low cost and because it works well on a broad spectrum of weeds common in the US corn belt, atrazine is commonly used with other herbicides to reduce the overall rate of atrazine and to lower the potential for groundwater contamination; it is a photosystem II inhibitor.

- Clopyralid is a broadleaf herbicide in the pyridine group, used mainly in turf, rangeland, and for control of noxious thistles. Notorious for its ability to persist in compost, it is another example of synthetic auxin.

- Dicamba, a postemergent broadleaf herbicide with some soil activity, is used on turf and field corn. It is another example of a synthetic auxin.

- Glufosinate ammonium, a broad-spectrum contact herbicide, is used to control weeds after the crop emerges or for total vegetation control on land not used for cultivation.

- Fluazifop (Fuselade Forte), a post emergence, foliar absorbed, translocated grass-selective herbicide with little residual action. It is used on a very wide range of broad leaved crops for control of annual and perennial grasses.

- Fluroxypyr, a systemic, selective herbicide, is used for the control of broad-leaved weeds in small grain cereals, maize, pastures, rangeland and turf. It is a synthetic auxin. In cereal growing, fluroxypyr's key importance is control of cleavers, *Galium aparine*. Other key broadleaf weeds are also controlled.

- Glyphosate, a systemic nonselective herbicide, is used in no-till burndown and for weed control in crops genetically modified to resist its effects. It is an example of an EPSPs inhibitor.

- Imazapyr a nonselective herbicide, is used for the control of a broad range of weeds, including terrestrial annual and perennial grasses and broadleaf herbs, woody species, and riparian and emergent aquatic species.

- Imazapic, a selective herbicide for both the pre- and postemergent control of some annual and perennial grasses and some broadleaf weeds, kills plants by inhibiting the production of branched chain amino acids (valine, leucine, and isoleucine), which are necessary for protein synthesis and cell growth.

- Imazamox, an imidazolinone manufactured by BASF for postemergence application that is an acetolactate synthase (ALS) inhibitor. Sold under trade names Raptor, Beyond, and Clearcast.

- Linuron is a nonselective herbicide used in the control of grasses and broadleaf weeds. It works by inhibiting photosynthesis.

- MCPA (2-methyl-4-chlorophenoxyacetic acid) is a phenoxy herbicide selective for broadleaf plants and widely used in cereals and pasture.

- Metolachlor is a pre-emergent herbicide widely used for control of annual grasses in corn and sorghum; it has displaced some of the atrazine in these uses.

- Paraquat is a nonselective contact herbicide used for no-till burndown and in aerial destruction of marijuana and coca plantings. It is more acutely toxic to people than any other herbicide in widespread commercial use.

- Pendimethalin, a pre-emergent herbicide, is widely used to control annual grasses and some broad-leaf weeds in a wide range of crops, including corn, soybeans, wheat, cotton, many tree and vine crops, and many turfgrass species.

- Picloram, a pyridine herbicide, mainly is used to control unwanted trees in pastures and edges of fields. It is another synthetic auxin.

- Sodium chlorate *(disused/banned in some countries)*, a nonselective herbicide, is considered phytotoxic to all green plant parts. It can also kill through root absorption.

- Triclopyr, a systemic, foliar herbicide in the pyridine group, is used to control broadleaf weeds while leaving grasses and conifers unaffected.

- Several sulfonylureas, including Flazasulfuron and Metsulfuron-methyl, which act as ALS inhibitors and in some cases are taken up from the soil via the roots.

Organic Herbicides

Recently, the term "organic" has come to imply products used in organic farming. Under this definition, an organic herbicide is one that can be used in a farming enterprise that has been classified as organic. Depending on the application, they may be less effective than synthetic herbicides and are generally used along with cultural and mechanical weed control practices.

Homemade organic herbicides include:

- Corn gluten meal (CGM) is a natural pre-emergence weed control used in turfgrass, which reduces germination of many broadleaf and grass weeds.

- Vinegar is effective for 5–20% solutions of acetic acid, with higher concentrations most effective, but it mainly destroys surface growth, so respraying to treat regrowth is needed. Resistant plants generally succumb when weakened by respraying.

- Steam has been applied commercially, but is now considered uneconomical and inadequate. It controls surface growth but not underground growth and so respraying to treat regrowth of perennials is needed.

- Flame is considered more effective than steam, but suffers from the same difficulties.

- D-limonene (citrus oil) is a natural degreasing agent that strips the waxy skin or cuticle from weeds, causing dehydration and ultimately death.

- Saltwater or salt applied in appropriate strengths to the rootzone will kill most plants.

Advantages and Disadvantages of Herbicides

The use of herbicides is increasing day by day. This is because the other alternative control measures do not provide an effective and economic substitute for herbicides in many situations. The efficacy and safety of herbicides are greatly influenced by soil and climate. These vary greatly between countries as does the legislation controlling their use.

Advantages of Herbicides on Weed Control

- They kill unwanted plants.

- They are easy to use.

- Herbicides can be used on closely planted crops where other methods cannot be used.

- Most of the time one application of the herbicide is enough whereas other methods have to be continually used.

- They work fast. They can be removed quickly in critical situations.

- Herbicides are relatively cheap, and most of the time cheaper than manual weeding.

On Crop Growth

- They can destroy plants bearing diseases.

- They help the crops grow by destroying the weed that causes harmful effects which include competition for water, nutrients and light; interference of weeds with crop growth by the release of toxins; modification of soil and air temperatures and the harbouring of pests.

- They can be safely used as the manual and mechanical removing of weeds can destroy the crop.

Others

- They are relatively safe on lands which may erode.

- Non-selective herbicides can effectively clear fields, where houses and roads can then be built.

Disadvantages of Herbicides

Effects of Herbicides on Environment

Herbicides vary greatly in chemical composition and in the degree of threat they pose to the environment. Many of the herbicides are highly persistant. It is widely recognised that the main reason accounting for residues of certain herbicides like simazine and other triazines in ground and surface water was the widespread use of these herbicides at high doses on hard surfaces.

- Soil: Some herbicides are non-biodegradable and are harmful for a long period of time. Heavy dose of herbicides affect microbial population of the soil. With herbicides targeting amino acid synthesis in both plants and microbes, there is a possibility that N2 fixation may be inhibited by the application of certain herbicides.

- Water: The improper use of pesticides and herbicides may also cause the storm water infiltration into groundwater. When these pesticides and herbicides contaminants dissolve in storm water they infiltrate the groundwater and then the surface waters, such as ponds, streams, rivers and lakes. These chemicals may also find their way into the soil and deeper groundwater units polluting them.

- Living organisms: Most herbicides are specifically plant poisons, and are not very toxic to animals. However, by changing the vegetation of treated sites, herbicide use also changes the habitat of birds, mammals, insects, and other animals through changes in the nature of their habitat. Herbivores may eat the plants treated with herbicides and then carnivores eat the herbivores. The toxic herbicide would be passed up the food chain increasing in concentration each time resulting in cancers and even deaths.

Anxiety about chemical residues in the environment has increased greatly in the last decade. These fears and concern about possible litigation have led many land managers to reappraise their weed control strategies. Change has also been forced on them by the decrease in the number of approved herbicides as a result of the high cost of registration. In addition, approval has been withdrawn from more toxic and persistent herbicides.

Effects of Herbicides on Humans

Among the many effects of pesticides and herbicides, perhaps the most alarming is the danger they pose to human health. People are directly affected by toxicity of some herbicides, during the course of their occupation (i.e., when spraying pesticides), or indirectly affected when exposed through drift or residues on food, and wildlife.

- Pesticides and herbicides can cause a number of health problems such as heart congestion, lung and kidney damage, low blood pressure, muscle damage, weight loss and adrenal glands damage.

- Arbitrary and indiscriminate usage of herbicides and pesticides can result in endometriosis, a common cause of infertility in women.

- Herbicides and pesticides have been suspected by the National Cancer Research Institute as a probable cause of certain cancers (i.e., cancers of the brain, prostrate, stomach and lip, as well as leukemia, skin melanomas, etc.) especially among farmers.

- The National Academy of Sciences reported that infants and children, because of their developing physiology, are susceptible to the negative effects of herbicides and pesticides in comparison to adults.

Effect of Herbicides On Crop Plant

An important problem with broadcast applications is that they are non-selective. They are toxic to a wide variety of plant species, and not just the weeds. If herbicides are not used properly, damage may be caused to crop plants, especially if too large dose is used, or if spraying occurs during a time when the crop species is sensitive to the herbicide. Unintended but economically important damage to crop plants is sometimes a consequence of the inappropriate use of herbicides.

Build-up of Resistant Biotypes

Apart from their effect on the environment, another major problem with herbicides has been the build-up of herbicide-resistant biotypes where the same herbicide has been used repeatedly for a number of years. This problem was not clearly foreseen at the start of the herbicide revolution but, since the early 1980s, triazine resistance has developed in most countries where these herbicides have been used. The usefulness of a number of other herbicides, including paraquat, dichlofopmethyl and sulfonylurea types has been affected by the development of resistant biotypes.

Methods of dealing with this problem include prevention of weed seed shedding, crop rotation, herbicide rotation, control of weed escapes and tillage practices. Crop rotation is not relevant in an amenity situation where the 'crops' are usually perennial but other control measures may be appropriate in certain situations. If weeds are prevented from setting seed, resistant biotypes cannot develop.

This could be achieved if land managers were made more aware of the threat of resistant biotypes and made greater efforts in intensively managed areas to prevent weeds from shedding seeds by the use of a rotation of herbicides supplemented by physical means such as mulching, hand hoeing and hand weeding.

Modern, intensively managed agricultural and forestry systems have an intrinsic reliance on the use of herbicides and other pesticides. Unfortunately, the use of herbicides and other pesticides carries risks to humans through exposure to these potentially toxic chemicals, and to ecosystems through direct toxicity caused to non-target species, and through changes in habitat. Nevertheless, until newer and more pest-specific solutions to weed-management problems are developed, there will be a continued reliance on herbicides in agriculture, forestry, and for other purposes, such as lawn care.

Crop Rotation for Weed Control

Ideally, weed management in an organic cropping system involves the integration of a broad range of cultural practices. Although cultivation after planting is usually a key component, a variety of other factors make important contributions to weed control on organic farms. All of these practices occur within the context of a sequence of crops that are planted on a field—the crop rotation. The identities of the crops are critical for disease and insect management, but for weed management the identity of the crop species is less critical than the type of soil preparation and cultural weed control practices used with each crop. In general, crop sequences that take advantage of multiple opportunities to suppress and remove weeds from the field will improve weed management on the farm.

Crop rotation in weed suppression, and most have looked at crop rotation in conventional agricultural systems with herbicides. the impact of varying the type of herbicide used rather than other factors associated with crop rotation. The few rotation studies without herbicides have generally found that more diverse systems had a lower density of problem weeds but a greater diversity of weed species. This is reasonable, since the variation in cultural practices during the rotation will tend to disrupt the life cycle of each particular weed species but create niches for a greater variety of species.

Although the sparse literature offers only general insights into the design of weed suppressive crop rotations, the ecology of weeds offers clues to the sort of crop sequences that are most likely to minimize weed problems. The usefulness of the principles outlined below has been documented by practicing farmers in many cases:

Include Clean Fallow Periods in the Rotation to Deplete Perennial Roots and Rhizomes and to Flush out and Destroy Annual Weeds.

Most perennial weed species will resprout after their roots and rhizomes have been cut into small pieces by tillage implements. Thus, many new shoots are produced, but each shoot is weakened due to less belowground food storage. If these plants are again cultivated into the soil, they are further weakened. A few such cycles once every two to three years can greatly suppress or eliminate most perennial weeds whose roots are within the plow layer and help keep deeper-rooted perennials in check. Similarly, tillage tends to stimulate the germination of seeds of most weed species, and subsequent cultivation will kill the seedlings so they will not compete with future crops. This is an important approach to reducing the density of weed seeds stored in the soil. In northern Pennsylvania, Eric and Anne Nordell have nearly eliminated weeds from their vegetable farm by alternating between a year with a cash crop and a year with weed-suppressing cover crops and a cultivated summer fallow. Over the years, they have greatly shortened the fallow periods as weeds have become less problematic.

Follow Weed-prone Crops with Crops in which Weeds can Easily be Prevented from going to Seed.

Weed control is inherently more difficult in some crops than others. For example, unless mulches are used, winter squash and pumpkins tend to become weedy because cultivation and hand weeding are essentially impossible after the vines have run out of the row. Moreover, weeds have usually set seed by the time these full-season crops have matured. Following such crops with a rapid succession of short-season crops like spinach and lettuce that are harvested before weeds can set seed will kill off many of the seeds produced in the vine crop. This reduces weed problems in subsequent crops. Similarly, small grains cannot be effectively harrowed after the stems begin to elongate, and, especially in spring-sown grain, weeds often go to seed before harvest. The long-term consequences of this seed production can be reduced by following the small grain with an easily cultivated, highly competitive crop like soybean or potato.

Weed control is often more difficult in direct-seeded vegetables than transplanted vegetables because the direct-seeded crops have a prolonged early period when the crop competes poorly and cultivation is difficult. Direct-seeded species like carrot and spinach that have small seeds are more of a problem in this regard than are large-seeded crops like snap beans and sweet corn. If weeds proliferate in the direct-seeded crop, it is usually advisable to follow it with an easily cultivated species that is transplanted or has a large seed.

Plant crops in which Weed Seed Production can be Prevented before Crops that are Poor Competitors

Weed control is often difficult in crops like onion and carrot because they are slow growing and cast relatively little shade. Although some weeds establish from the long-term seed bank in the soil, many of the weeds encountered in a given year establish from seeds shed in the previous year or two. Consequently, growing a crop in which weed seed production can be prevented before planting a poor competitor can reduce the amount of precision cultivation and hand weeding required for successful production of the poor competitor. Cropping strategies in which management prevents weed seed production include successive plantings of short-season crops, short-cycle cover crops alternating with clean fallow periods, crops grown with weed-suppressing mulch, and highly competitive crops that are intensively cultivated (for example, potato). The crop types that prevent seed production will vary depending on the practices of the farmer. For example, mulched vine crops may be used as a weed "cleaning" crop because the combination of mulch and a dense canopy effectively suppresses weeds, whereas without mulch, vine crops often contribute to the weed seed bank because they become difficult to cultivate or hand weed late in the season.

If weeds become a serious barrier to the production of noncompetitive crops, growing poor competitors in a special crop rotation in which they alternate only with cleaning crops may prove worthwhile. When weed populations have declined substantially, the rotation can be broadened to include occasional crops in which seed production is more difficult to prevent.

Rotate between Crops that are Planted in Different Seasons

Weed species have characteristic times of the year during which they emerge. Common ragweed emerges most readily in early spring and is often a problem in spring-sown organic small grains like barley and oat. In contrast, henbit and shepherd's purse are typically fall-germinating species and are likely to be found in winter wheat and spelt. Rotations that include both spring-planted and fall-planted crops tend to suppress both sorts of species. The spring-germinating weeds tend to be competitively suppressed by fall-sown grain because it is already well established and growing vigorously by the time they germinate. Conversely, fall-germinating species tend to be destroyed before they can set seed when soil is tilled for a spring crop.

Similarly, competitive spring-planted crops tend to suppress midsummer germinating species like purslane, whereas summer tillage for midseason planted vegetables will kill most spring-germinating weed species before they can set seeds.

Work Cover Crops into the Rotation between Cash Crops at Times when the Soil would otherwise be Bare

Weeds establish most easily when the ground is bare. Plant canopies suppress seed germination of many weed species by reducing the amount of light and the relative amount of red-wavelength light reaching the soil surface. In addition, cover crops compete with any weeds that do emerge. Thus, for example, planting winter rye or mustard cover crops following plow-down of pasture reduced spring weed cover from 52 percent to 9 percent with rye and to 4 percent with mustard.

That Grain and legume crops are more competitive against weeds when planted at high density and uniformity, and the same principle applies when these species are used as cover crops. Increasing seeding rates by 50 to 100 percent relative to recommended rates for grain or forage production usually produces noticeably improved weed control, particularly if the cover crop is broadcast. Very high-density sowings of competitive cover crops like buckwheat, soybean, and grain rye can completely smother even many perennial weeds. Experimenting on small areas to find the right balance between seed cost of cover crops and weed control is often worthwhile.

Cover crops interseeded into standing cash crops during the last cultivation can help suppress late-emerging weeds but may also compete with the crop.

Avoid Cover Crop Species and Cover Crop Management that Promote Weeds

Many cover crops can behave as weeds if allowed to go to seed due to a delay in mowing or incorporation. Buckwheat and winter grains are particularly prone to cause problems if allowed to seed, because they are fast-growing, competitive species.

Hairy vetch should be avoided on any farm that grows winter grain. Some portion of any hairy vetch seed lot will be "hard" seeds (seeds that are dormant due to a seed coat that does not allow absorption of water). Even if the hairy vetch cover crop is not allowed to go to seed, some of the hard seeds from the original sowing will germinate in subsequent winter grain crops and can severely reduce both yield and grain quality.

Long-season cover crops like red clover and sweet clover are useful for supplying nitrogen and improving soil structure, but they are relatively uncompetitive early in their development and can become weed infested. Because they remain in the ground most of a year or more, annual weeds have time to set seeds, and perennials have plenty of time to increase by growth of roots and rhizomes (underground stems). Such cover crops thus work best when sown with or interseeded into a grain "nurse" crop that can suppress weeds while the legume cover crop develops. Even with a nurse crop, however, weeds can be a problem in the following crop, and many growers choose to avoid red clover and sweet clover before most vegetable crops.

Even hairy vetch or a winter grain like rye may allow seed production by winter annuals like chickweed and shepherd's purse if the cover crop is not incorporated promptly in the spring. The problem is greatest when the cover crop stand is light or spotty. The problem can be avoided, however, by incorporating the cover crop at the first sign of capsule formation on the weeds.

Rotate between Annual Crops and Perennial Sod Crops

Although weed seed populations decline more rapidly when the soil is tilled, substantial decreases in populations of most annual weeds occur when sod crops are left in the ground for a few years. This occurs by natural die-off of seeds in the soil and because annuals that germinate in a perennial sod are competitively suppressed by the already well-established perennial legumes and grasses. The few annuals that do establish are usually prevented from setting seeds by repeated mowing or grazing. One study found that 83 percent of farmers surveyed in Saskatchewan and Manitoba noticed decreased weed problems following sod crops, and most indicated that the effect lasted more than one year. In addition to reducing annual weeds, many of these farmers indicated a reduction in Canada thistle, probably because mowing several times each year depleted food storage in the thistle's roots.

Precautions are required, however, when alternating sod crops with annual crops. Perennial grass weeds like quackgrass should be well controlled prior to planting the sod, or they will likely increase during the sod phase of the rotation. Also, annual weeds may set many seeds during establishment of the sod crop, thereby negating the expected decline in the weed seed bank. Weed seed production can be minimized by using a grain nurse crop that competes with the annual weeds, early harvest of the nurse crop for forage or straw before annuals go to seed, and subsequent mowing of the sod during the establishment year. Another strategy to reduce weed seed production during sod establishment is to plant the sod crop in late summer or early fall.

Although the impact of crop rotation on weeds will not be seen as quickly as the impact of tillage or cultivation, the cumulative affect of a well-planned rotation strategy can, over several years, greatly decrease weed density. Rotation planning is a key way organic growers can substitute brain power for labor and purchased inputs. This principle applies not just to weeds, but also to diseases, insects, soil nutrients and soil health.

References

- "Managing Invasive Plants: Concepts, Principles, and Practices". United States Fish & Wildlife Service. Retrieved 19 Feb 2012

- Weed-control, terms: sciencedaily.com, Retrieved 6 June, 2019

- Turgeon; et al. (2009). Weed Control in Turf and Ornamentals. Upper Saddle River, New Jersey: Pearson Education, Inc. Pp. 127–129. ISBN 0-13-159122-3

- Weed-management, forest-protection, forestry, agriculture: vikaspedia.in, Retrieved 7 July, 2019

- Zhou Q, Liu W, Zhang Y, Liu KK (Oct 2007). "Action mechanisms of acetolactate synthase-inhibiting herbicides". Pesticide Biochemistry and Physiology. 89 (2): 89–96. Doi:10.1016/j.pestbp.2007.04.004

- What-is-integrated-weed-management, iwm-toolbox: integratedweedmanagement.org, Retrieved 8 August, 2019

- Powles, S. B.; Shaner, D. L., eds. (2001). Herbicide Resistance and World Grains. CRC Press, Boca Raton, FL. P. 328. ISBN 9781420039085

- Overview-of-biological-methods-of-weed-control, biological-approaches-for-controlling-weeds: intechopen.com, Retrieved 9 January, 2019

- Environmental Protection Agency: Atrazine Updates.Current as of January 2013. Retrieved August 24, 2013

- Chemical-methods-weed-control-option-weed-management: morungexpress.com, Retrieved 10 February, 2019

4

Pest Control Methods

Pest control is the management and regulation of pests which are harmful to humans. It can be categorized into biological, physical, mechanical and chemical methods of pest control including fungicides, trap crop, rodenticides, miticides, etc. This chapter discusses these methods of pest control in detail.

Pest Control

Pest control is a huge category in the horticulture industry. Essentially, pest control encompasses everything a gardener does both to prevent pests from attacking their plants, and eradicate pest infestations once they occur.

In horticulture, especially in greenhouses and grow rooms, pest control options include chemical and organic pesticides, as well as a host of other options such as integrated pest management (IPM), sticky traps, screens, good personal hygiene, environmental control, and more.

There are several branches of pest control, including biological pest control, which is pitting nature against nature. In these cases, beneficial insects are unleashed in a confined space to attack (eat) the pest insects.

Indeed, pests can reduce food security, having a severe impact on yields. While pesticides are commonly used to treat various types of pests, these substances have also been shown to be quite harmful to plants, in some cases even killing them. An abuse (overuse) of pesticides can even pollute water supplies while having a negative impact on the overall health of the farmers.

For the above reasons, some farmers opt for alternative pest control methods such as crop rotation. This involves occasionally alternating between different crop species in order to stop the pests from getting used to the same types of plants. This method also has the advantage of encouraging soil fertility through nitrogen diffused from the roots.

Alternatively, some planters also choose to go for polyculture instead of pesticides. This entails growing different types of crops in the same space, which has been known to increase local biodiversity while reducing the crop's susceptibility to different types of pests and diseases. On the flip side, polyculture does require more manpower and maintenance.

Other viable pest control options include setting sticky traps up, or planting banker plants, which, in addition to polyculture, is a form of companion planting.

Biological Pest Control

Syrphus hoverfly larva (below) feed on aphids (above), making them natural biological control agents.

A parasitoid wasp (*Cotesia congregata*) adult with pupal cocoons on its host, a tobacco hornworm (*Manduca sexta*, green background), an example of a hymenopteran biological control agent.

Biological control or biocontrol is a method of controlling pests such as insects, mites, weeds and plant diseases using other organisms. It relies on predation, parasitism, herbivory, or other natural mechanisms, but typically also involves an active human management role. It can be an important component of integrated pest management (IPM) programs.

There are three basic strategies for biological pest control: classical (importation), where a natural enemy of a pest is introduced in the hope of achieving control; inductive (augmentation), in which a large population of natural enemies are administered for quick pest control; and inoculative (conservation), in which measures are taken to maintain natural enemies through regular reestablishment.

Natural enemies of insect pests, also known as biological control agents, include predators, parasitoids, pathogens, and competitors. Biological control agents of plant diseases are most often referred to as antagonists. Biological control agents of weeds include seed predators, herbivores and plant pathogens.

Biological control can have side-effects on biodiversity through attacks on non-target species by any of the same mechanisms, especially when a species is introduced without thorough understanding of the possible consequences.

Types of Biological Pest Control

There are three basic biological pest control strategies: importation (classical biological control), augmentation and conservation.

Importation

Rodolia cardinalis, the vedalia beetle, was imported from Australia to California in the 19th century, successfully controlling cottony cushion scale.

Importation or classical biological control involves the introduction of a pest's natural enemies to a new locale where they do not occur naturally. Early instances were often unofficial and not based on research, and some introduced species became serious pests themselves.

To be most effective at controlling a pest, a biological control agent requires a colonizing ability which allows it to keep pace with changes to the habitat in space and time. Control is greatest if the agent has temporal persistence, so that it can maintain its population even in the temporary absence of the target species, and if it is an opportunistic forager, enabling it to rapidly exploit a pest population.

Joseph Needham noted a Chinese text dating from 304 AD, *Records of the Plants and Trees of the Southern Regions*, by Hsi Han, which describes mandarin oranges protected by large reddish-yellow citrus ants which attack and kill insect pests of the orange trees. The citrus ant (*Oecophylla smaragdina*) was rediscovered in the 20th century, and since 1958 has been used in China to protect orange groves.

One of the earliest successes in the west was in controlling *Icerya purchasi* (cottony cushion scale) in Australia, using a predatory insect *Rodolia cardinalis* (the vedalia beetle). This success was repeated in California using the beetle and a parasitoidal fly, *Cryptochaetum iceryae*. Other successful cases include the control of *Antonina graminis* in Texas by *Neodusmetia sangwani* in the 1960s.

Damage from *Hypera postica*, the alfalfa weevil, a serious introduced pest of forage, was substantially reduced by the introduction of natural enemies. 20 years after their introduction the population of weevils in the alfalfa area treated for alfalfa weevil in the Northeastern United States remained 75 percent down.

The invasive species *Alternanthera philoxeroides* (alligator weed) was controlled in Florida (U.S.) by introducing alligator weed flea beetle.

Alligator weed was introduced to the United States from South America. It takes root in shallow water, interfering with navigation, irrigation, and flood control. The alligator weed flea beetle and two other biological controls were released in Florida, greatly reducing the amount of land covered by the plant. Another aquatic weed, the giant salvinia (*Salvinia molesta*) is a serious pest, covering waterways, reducing water flow and harming native species. Control with the salvinia weevil (*Cyrtobagous salviniae*) and the salvinia stem-borer moth (*Samea multiplicalis)* is effective in warm climates, and in Zimbabwe, a 99% control of the weed was obtained over a two-year period.

Small commercially reared parasitoidal wasps, *Trichogramma ostriniae*, provide limited and erratic control of the European corn borer (*Ostrinia nubilalis*), a serious pest. Careful formulations of the bacterium *Bacillus thuringiensis* are more effective.

The population of *Levuana iridescens*, the Levuana moth, a serious coconut pest in Fiji, was brought under control by a classical biological control program in the 1920s.

Augmentation

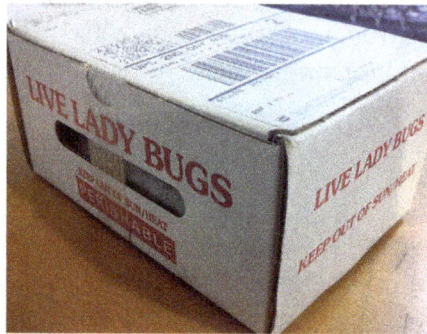

Hippodamia convergens, the convergent lady beetle, is commonly
sold for biological control of aphids.

Augmentation involves the supplemental release of natural enemies that occur in a particular area, boosting the naturally occurring populations there. In inoculative release,
small numbers of the control agents are released at intervals to allow them to reproduce, in the hope of setting up longer-term control, and thus keeping the pest down to
a low level, constituting prevention rather than cure. In inundative release, in contrast,
large numbers are released in the hope of rapidly reducing a damaging pest population,
correcting a problem that has already arisen. Augmentation can be effective, but is not
guaranteed to work, and depends on the precise details of the interactions between
each pest and control agent.

An example of inoculative release occurs in the horticultural production of several
crops in greenhouses. Periodic releases of the parasitoidal wasp, *Encarsia formosa*, are
used to control greenhouse whitefly, while the predatory mite *Phytoseiulus persimilis*
is used for control of the two-spotted spider mite.

The egg parasite *Trichogramma* is frequently released inundatively to control harmful
moths. Similarly, *Bacillus thuringiensis* and other microbial insecticides are used in
large enough quantities for a rapid effect. Recommended release rates for *Trichogramma* in vegetable or field crops range from 5,000 to 200,000 per acre (1 to 50 per square
metre) per week according to the level of pest infestation. Similarly, nematodes that kill
insects (that are entomopathogenic) are released at rates of millions and even billions
per acre for control of certain soil-dwelling insect pests.

Conservation

The conservation of existing natural enemies in an environment is the third method
of biological pest control. Natural enemies are already adapted to the habitat and to
the target pest, and their conservation can be simple and cost-effective, as when nec-
tar-producing crop plants are grown in the borders of rice fields. These provide nectar
to support parasitoids and predators of planthopper pests and have been demonstrated
to be so effective (reducing pest densities by 10- or even 100-fold) that farmers sprayed

70% less insecticides and enjoyed yields boosted by 5%. Predators of aphids were similarly found to be present in tussock grasses by field boundary hedges in England, but they spread too slowly to reach the centres of fields. Control was improved by planting a metre-wide strip of tussock grasses in field centres, enabling aphid predators to overwinter there.

An inverted flowerpot filled with straw to attract earwigs.

Cropping systems can be modified to favor natural enemies, a practice sometimes referred to as habitat manipulation. Providing a suitable habitat, such as a shelterbelt, hedgerow, or beetle bank where beneficial insects such as parasitoidal wasps can live and reproduce, can help ensure the survival of populations of natural enemies. Things as simple as leaving a layer of fallen leaves or mulch in place provides a suitable food source for worms and provides a shelter for insects, in turn being a food source for such beneficial mammals as hedgehogs and shrews. Compost piles and stacks of wood can provide shelter for invertebrates and small mammals. Long grass and ponds support amphibians. Not removing dead annuals and non-hardy plants in the autumn allows insects to make use of their hollow stems during winter. In California, prune trees are sometimes planted in grape vineyards to provide an improved overwintering habitat or refuge for a key grape pest parasitoid. The providing of artificial shelters in the form of wooden caskets, boxes or flowerpots is also sometimes undertaken, particularly in gardens, to make a cropped area more attractive to natural enemies. For example, earwigs are natural predators which can be encouraged in gardens by hanging upside-down flowerpots filled with straw or wood wool. Green lacewings can be encouraged by using plastic bottles with an open bottom and a roll of cardboard inside. Birdhouses enable insectivorous birds to nest; the most useful birds can be attracted by choosing an opening just large enough for the desired species.

In cotton production, the replacement of broad-spectrum insecticides with selective control measures such as Bt cotton can create a more favorable environment for natural enemies of cotton pests due to reduced insecticide exposure risk. Such predators or parasitoids can control pests not affected by the Bt protein. Reduced prey quality and abundance associated increased control from Bt cotton can also indirectly decrease natural enemy populations in some cases, but the percentage of pests eaten or parasitized in Bt and non-Bt cotton are often similar.

Biological Control Agents

Predators

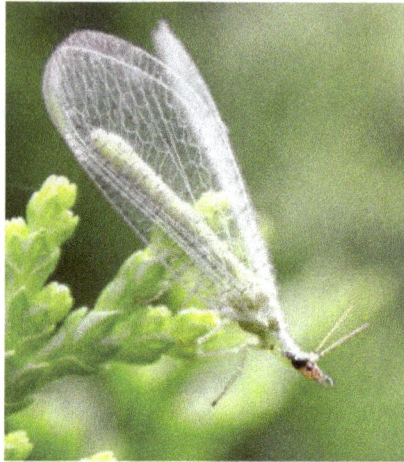

Predatory lacewings are available from biocontrol dealers.

Predators are mainly free-living species that directly consume a large number of prey during their whole lifetime. Given that many major crop pests are insects, many of the predators used in biological control are insectivorous species. Lady beetles, and in particular their larvae which are active between May and July in the northern hemisphere, are voracious predators of aphids, and also consume mites, scale insects and small caterpillars. The spotted lady beetle (*Coleomegilla maculata*) is also able to feed on the eggs and larvae of the Colorado potato beetle (*Leptinotarsa decemlineata*).

The larvae of many hoverfly species principally feed upon aphids, one larva devouring up to 400 in its lifetime. Their effectiveness in commercial crops has not been studied.

Predatory *Polistes* wasp searching for bollworms or other caterpillars on a cotton plant.

Several species of entomopathogenic nematode are important predators of insect and other invertebrate pests. *Phasmarhabditis hermaphrodita* is a microscopic nematode that kills slugs. Its complex life cycle includes a free-living, infective stage in the soil where it becomes associated with a pathogenic bacteria such as *Moraxella osloensis*. The nematode enters the slug through the posterior mantle region, thereafter feeding and reproducing inside, but it is the bacteria that kill the slug. The nematode is available commercially in Europe and is applied by watering onto moist soil.

Species used to control spider mites include the predatory mites *Phytoseiulus persimilis, Neoseilus californicus,* and *Amblyseius cucumeris*, the predatory midge *Feltiella acarisuga*, and a ladybird *Stethorus punctillum*. The bug *Orius insidiosus* has been successfully used against the two-spotted spider mite and the western flower thrips (*Frankliniella occidentalis*).

Predators including *Cactoblastis cactorum* (mentioned above) can also be used to destroy invasive plant species. As another example, the poison hemlock moth (*Agonopterix alstroemeriana)* can be used to control poison hemlock (*Conium maculatum*). During its larval stage, the moth strictly consumes its host plant, poison hemlock, and can exist at hundreds of larvae per individual host plant, destroying large swathes of the hemlock.

The parasitoid wasp *Aleiodes indiscretus* parasitizing a gypsy moth caterpillar, a serious pest of forestry.

For rodent pests, cats are effective biological control when used in conjunction with reduction of "harborage"/hiding locations. While cats are effective at preventing rodent "population explosions", they are not effective for eliminating pre-existing severe infestations. Barn owls are also sometimes used as biological rodent control. Although there are no quantitative studies of the effectiveness of barn owls for this purpose, they are known rodent predators that can be used in addition to or instead of cats; they can be encouraged into an area with nest boxes.

Parasitoids

Parasitoids lay their eggs on or in the body of an insect host, which is then used as a food for developing larvae. The host is ultimately killed. Most insect parasitoids are wasps or flies, and many have a very narrow host range. The most important groups are the ichneumonid wasps, which mainly use caterpillars as hosts; braconid wasps, which attack caterpillars and a wide range of other insects including aphids; chalcid wasps, which parasitize eggs and larvae of many insect species; and tachinid flies, which parasitize a wide range of insects including caterpillars, beetle adults and larvae, and true bugs. Parasitoids are most effective at reducing pest populations when their host organisms have limited refuges to hide from them.

Encarsia formosa, widely used in greenhouse horticulture, was one of the first biological control agents developed.

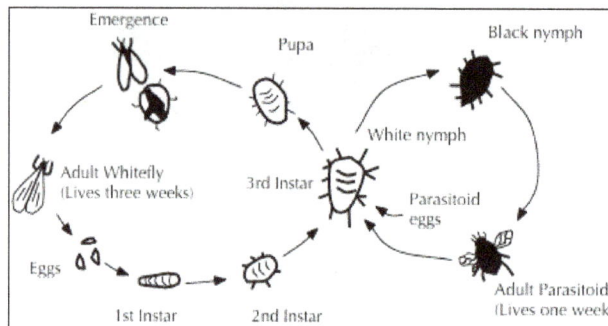

Life cycles of greenhouse whitefly and its parasitoid wasp *Encarsia formosa*.

Parasitoids are among the most widely used biological control agents. Commercially, there are two types of rearing systems: short-term daily output with high production of parasitoids per day, and long-term, low daily output systems. In most instances, production will need to be matched with the appropriate release dates when susceptible host species at a suitable phase of development will be available. Larger production facilities produce on a yearlong basis, whereas some facilities produce only seasonally. Rearing facilities are usually a significant distance from where the agents are to be used in the field, and transporting the parasitoids from the point of production to the point of use can pose problems. Shipping conditions can be too hot, and even vibrations from planes or trucks can adversely affect parasitoids.

Encarsia formosa is a small predatory chalcid wasp which is a parasitoid of white-fly, a sap-feeding insect which can cause wilting and black sooty moulds in glasshouse vegetable and ornamental crops. It is most effective when dealing with low level infestations, giving protection over a long period of time. The wasp lays its eggs in young whitefly 'scales', turning them black as the parasite larvae pupate. *Gonatocerus ashmeadi* (Hymenoptera: Mymaridae) has been introduced to control the glassy-winged sharpshooter *Homalodisca vitripennis* (Hemiptera: Cicadellidae) in French Polynesia and has successfully controlled ~95% of the pest density.

The eastern spruce budworm is an example of a destructive insect in fir and spruce forests. Birds are a natural form of biological control, but the *Trichogramma minutum*, a species of parasitic wasp, has been investigated as an alternative to more controversial chemical controls.

There are a number of recent studies pursuing sustainable methods for controlling urban cockroaches using parasitic wasps. Since most cockroaches remain in the sewer system and sheltered areas which are inaccessible to insecticides, employing active-hunter wasps is a strategy to try and reduce their populations.

Pathogens

Pathogenic micro-organisms include bacteria, fungi, and viruses. They kill or debilitate their host and are relatively host-specific. Various microbial insect diseases occur naturally, but may also be used as biological pesticides. When naturally occurring, these outbreaks are density-dependent in that they generally only occur as insect populations become denser.

Bacteria

Bacteria used for biological control infect insects via their digestive tracts, so they offer only limited options for controlling insects with sucking mouth parts such as aphids and scale insects. *Bacillus thuringiensis*, a soil-dwelling bacterium, is the most widely applied species of bacteria used for biological control, with at least four sub-species used against Lepidopteran (moth, butterfly), Coleopteran (beetle) and Dipteran (true fly) insect pests. The bacterium is available to organic farmers in sachets of dried spores which are mixed with water and sprayed onto vulnerable plants such as brassicas and fruit trees. Genes from *B. thuringiensis* have also been incorporated into transgenic crops, making the plants express some of the bacterium's toxins, which are proteins. These confer resistance to insect pests and thus reduce the necessity for pesticide use. If pests develop resistance to the toxins in these crops, *B. thuringiensis* will become useless in organic farming also. The bacterium *Paenibacillus popilliae* which causes milky spore disease has been found useful in the control of Japanese beetle, killing the larvae. It is very specific to its host species and is harmless to vertebrates and other invertebrates.

Fungi

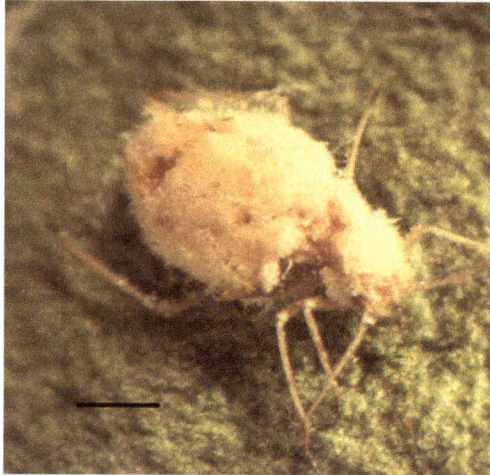

Green peach aphid, a pest in its own right and a vector of plant viruses, killed
by the fungus *Pandora neoaphidis* (Zygomycota: Entomophthorales) Scale bar = 0.3 mm.

Entomopathogenic fungi, which cause disease in insects, include at least 14 species that attack aphids. *Beauveria bassiana* is mass-produced and used to manage a wide variety of insect pests including whiteflies, thrips, aphids and weevils. *Lecanicillium* spp. are deployed against white flies, thrips and aphids. *Metarhizium* spp. are used against pests including beetles, locusts and other grasshoppers, Hemiptera, and spider mites. *Paecilomyces fumosoroseus* is effective against white flies, thrips and aphids; *Purpureocillium lilacinus* is used against root-knot nematodes, and 89 *Trichoderma* species against certain plant pathogens. *Trichoderma viride* has been used against Dutch elm disease, and has shown some effect in suppressing silver leaf, a disease of stone fruits caused by the pathogenic fungus *Chondrostereum purpureum*.

The fungi *Cordyceps* and *Metacordyceps* are deployed against a wide spectrum of arthropods. *Entomophaga* is effective against pests such as the green peach aphid.

Several members of Chytridiomycota and Blastocladiomycota have been explored as agents of biological control. From Chytridiomycota, *Synchytrium solstitiale* is being considered as a control agent of the yellow star thistle (*Centaurea solstitialis*) in the United States.

Viruses

Baculoviruses are specific to individual insect host species and have been shown to be useful in biological pest control. For example, the Lymantria dispar multicapsid nuclear polyhedrosis virus has been used to spray large areas of forest in North America where larvae of the gypsy moth are causing serious defoliation. The moth larvae are killed by the virus they have eaten and die, the disintegrating cadavers leaving virus particles on the foliage to infect other larvae.

A mammalian virus, the rabbit haemorrhagic disease virus was introduced to Australia to attempt to control the European rabbit populations there. It escaped from quarantine and spread across the country, killing large numbers of rabbits. Very young animals survived, passing immunity to their offspring in due course and eventually producing a virus-resistant population. Introduction into New Zealand in the 1990s was similarly successful at first, but a decade later, immunity had developed and populations had returned to pre-RHD levels.

Oomycota

Lagenidium giganteum is a water-borne mold that parasitizes the larval stage of mosquitoes. When applied to water, the motile spores avoid unsuitable host species and search out suitable mosquito larval hosts. This mold has the advantages of a dormant phase, resistant to desiccation, with slow-release characteristics over several years. Unfortunately, it is susceptible to many chemicals used in mosquito abatement programmes.

Competitors

The legume vine *Mucuna pruriens* is used in the countries of Benin and Vietnam as a biological control for problematic *Imperata cylindrica* grass: the vine is extremely vigorous and suppresses neighbouring plants by out-competing them for space and light. *Mucuna pruriens* is said not to be invasive outside its cultivated area. *Desmodium uncinatum* can be used in push-pull farming to stop the parasitic plant, witchweed (*Striga*).

The Australian bush fly, *Musca vetustissima*, is a major nuisance pest in Australia, but native decomposers found in Australia are not adapted to feeding on cow dung, which is where bush flies breed. Therefore, the Australian Dung Beetle Project (1965–1985), led by George Bornemissza of the Commonwealth Scientific and Industrial Research Organisation, released forty-nine species of dung beetle, to reduce the amount of dung and therefore also the potential breeding sites of the fly.

Combined use of Parasitoids and Pathogens

In cases of massive and severe infection of invasive pests, techniques of pest control are often used in combination. An example is the emerald ash borer, *Agrilus planipennis*, an invasive beetle from China, which has destroyed tens of millions of ash trees in its introduced range in North America. As part of the campaign against it, from 2003 American scientists and the Chinese Academy of Forestry searched for its natural enemies in the wild, leading to the discovery of several parasitoid wasps, namely *Tetrastichus planipennisi*, a gregarious larval endoparasitoid, *Oobius agrili*, a solitary, parthenogenic egg parasitoid, and *Spathius agrili*, a gregarious larval ectoparasitoid. These have been introduced and released into the United States of America as a possible biological control of the emerald ash borer. Initial results for *Tetrastichus planipennisi* have shown promise, and it is now being released along with *Beauveria bassiana*, a fungal pathogen with known insecticidal properties.

Difficulties

Many of the most important pests are exotic, invasive species that severely impact agriculture, horticulture, forestry and urban environments. They tend to arrive without their co-evolved parasites, pathogens and predators, and by escaping from these, populations may soar. Importing the natural enemies of these pests may seem a logical move but this may have unintended consequences; regulations may be ineffective and there may be unanticipated effects on biodiversity, and the adoption of the techniques may prove challenging because of a lack of knowledge among farmers and growers.

Side Effects

Biological control can affect biodiversity through predation, parasitism, pathogenicity, competition, or other attacks on non-target species. An introduced control does not always target only the intended pest species; it can also target native species. In Hawaii during the 1940s parasitic wasps were introduced to control a lepidopteran pest and the wasps are still found there today. This may have a negative impact on the native ecosystem; however, host range and impacts need to be studied before declaring their impact on the environment.

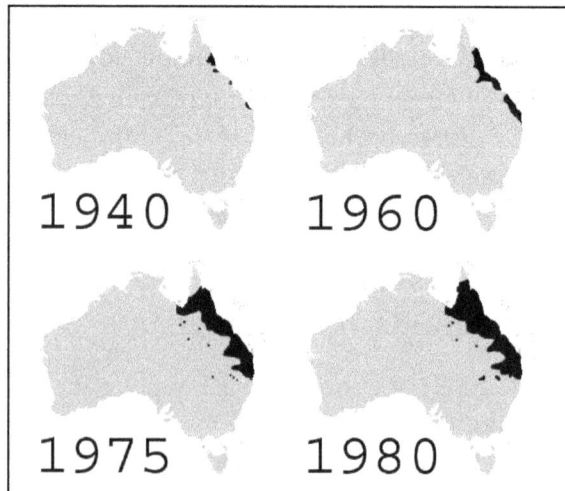

Cane toad (introduced into Australia 1935) spread from 1940 to 1980:
it was ineffective as a control agent. Its distribution has continued to widen since 1980.

Vertebrate animals tend to be generalist feeders, and seldom make good biological control agents; many of the classic cases of "biocontrol gone awry" involve vertebrates. For example, the cane toad (*Rhinella marina*) was intentionally introduced to Australia to control the greyback cane beetle (*Dermolepida albohirtum*), and other pests of sugar cane. 102 toads were obtained from Hawaii and bred in captivity to increase their numbers until they were released into the sugar cane fields of the tropic north in 1935. It was later discovered that the toads could not jump very high and so were unable to eat the cane beetles which stayed on the upper stalks of the cane plants. However, the toad thrived by feeding on other insects and soon spread very rapidly; it took over native amphibian habitat and brought foreign disease to native toads and frogs, dramatically reducing their populations. Also, when it is threatened or handled, the cane toad releases poison from parotoid glands on its shoulders; native Australian species such as goannas, tiger snakes, dingos and northern quolls that attempted to eat the toad were harmed or killed. However, there has been some recent evidence that native predators are adapting, both physiologically and through changing their behaviour, so in the long run, their populations may recover.

Rhinocyllus conicus, a seed-feeding weevil, was introduced to North America to control exotic musk thistle (*Carduus nutans*) and Canadian thistle (*Cirsium arvense*). However, the weevil also attacks native thistles, harming such species as the endemic Platte thistle (*Cirsium neomexicanum*) by selecting larger plants (which reduced the gene pool), reducing seed production and ultimately threatening the species' survival. Similarly, the weevil *Larinus planus* was also used to try to control the Canadian thistle, but it damaged other thistles as well. This included one species classified as threatened.

The small Asian mongoose (*Herpestus javanicus*) was introduced to Hawaii in order to control the rat population. However, the mongoose was diurnal, and the rats emerged at night; the mongoose therefore preyed on the endemic birds of Hawaii, especially their eggs, more often than it ate the rats, and now both rats and mongooses threaten the birds. This introduction was undertaken without understanding the consequences of such an action. No regulations existed at the time, and more careful evaluation should prevent such releases now.

The sturdy and prolific eastern mosquitofish (*Gambusia holbrooki*) is a native of the southeastern United States and was introduced around the world in the 1930s and '40s to feed on mosquito larvae and thus combat malaria. However, it has thrived at the expense of local species, causing a decline of endemic fish and frogs through competition for food resources, as well as through eating their eggs and larvae. In Australia, control of the mosquitofish is the subject of discussion; in 1989 researchers A. H. Arthington and L. L. Lloyd stated that "biological population control is well beyond present capabilities".

Grower Education

A potential obstacle to the adoption of biological pest control measures is that growers may prefer to stay with the familiar use of pesticides. However, pesticides have undesired effects, including the development of resistance among pests, and the destruction of natural enemies; these may in turn enable outbreaks of pests of other species than the ones originally targeted, and on crops at a distance from those treated with pesticides. One method of increasing grower adoption of biocontrol methods involves letting them learn by doing, for example showing them simple field experiments, enabling them to observe the live predation of pests, or demonstrations of parasitised pests. In the Philippines, early season sprays against leaf folder caterpillars were common practice, but growers were asked to follow a 'rule of thumb' of not spraying against leaf folders for the first 30 days after transplanting; participation in this resulted in a reduction of insecticide use by 1/3 and a change in grower perception of insecticide use.

Trap Crop

A trap crop is a plant that attracts agricultural pests, usually insects, away from nearby crops. This form of companion planting can save the main crop from decimation by pests without the use of pesticides. While many trap crops have successfully diverted pests off of focal crops in small scale greenhouse, garden and field experiments, only a small portion of these plants have been shown to reduce pest damage at larger commercial scales. A common explanation for reported trap cropping failures, is that attractive trap plants only protect nearby plants if the insects do not move back into the main crop. In a review of 100 trap cropping examples in 2006, only 10 trap crops were classified as successful at a commercial scale, and in all successful cases, trap cropping was supplemented with management practices that specifically limited insect dispersal from the trap crop back into the main crop.

Usage

Trap crops, when used on an industrial scale, are generally planted at a key time in the pest's life-cycle, and then destroyed before that life-cycle finishes and the pest might have transferred from the trap plants to the main crop.

Examples of trap crops include:

- Alfalfa planted in strips among cotton, to draw away lygus bugs, while castor beans surround the field, or tobacco is planted in strips among it, to protect from the budworm *Heliothis*.

- Rose enthusiasts often plant *Pelargonium* geraniums among their rosebushes because Japanese beetles are drawn to the geraniums, which are toxic to them.

- Chervil is used by gardeners to protect vegetable plants from slugs.

- Rye, sesbania, and sicklepod are used to protect soybeans from corn seeding maggots, stink bugs, and velvet green caterpillars, respectively.

- Mustard and Alfalfa planted near strawberries to attract lygus bugs, a method pioneered by Jim Cochran.

Trap crops can be planted around the circumference of the field to be protected, which is assumed to act as a barrier for entry by pests, or they can be interspersed among the focul crop, for example being planted every ninth row. Planting crops in rows helps facilitate supplemental management practices that prevent insect pest dispersal back into the main field, such as driving a vehicle above the trap crop which then removes insect pests by vacuuming them off of the trap crop row or targeted insecticides, which are only deployed on the trap crop. Even if pesticides are used to control insects on the trap crop, total pesticides are greatly reduced in this scenario over conventional agricultural pesticide applications because they are only deployed on a small portion of the farm (the trap crop). Other strategies that prevent dispersal of insect pests back into the main crop include cutting the trap plants, applying predators to the trap plant that eat the pest, and planting a high ratio of trap plants to other plants.

Operation

Recent studies on host-plant finding have shown that flying pests are far less successful if their host-plants are surrounded by any other plant, or even "decoy-plants" made of green plastic, cardboard or any other green material. The host-plant finding process occurs in three phases.

The first phase is stimulation by odours characteristic to the host-plant. This induces the insect to try to land on the plant it seeks. But insects avoid landing on brown (bare) soil. So if only the host-plant is present, the insects will quasi-systematically find it by landing on the only green thing around. This is called an "appropriate landing". When it does an "inappropriate landing", it flies off to any other nearby patch of green. It eventually leaves the area if there are too many "inappropriate" landings.

The second phase of host-plant finding is for the insect to make short flights from leaf to leaf to assess the plant's overall suitability. The number of leaf-to-leaf flights varies according to the insect species and to the host-plant stimulus received from each leaf. But the insect must accumulate sufficient stimuli from the host-plant to lay eggs; so it must make a certain number of consecutive "appropriate" landings. Hence if it makes an "inappropriate landing", the assessment of that plant is negative and the insect must start the process anew.

Thus it was shown that clover used as a ground cover had the same disruptive effect on eight pest species from four insect orders. An experiment showed that 36% of cabbage root flies laid eggs beside cabbages growing in bare soil (which resulted in no crop), compared with only 7% beside cabbages growing in clover (which allowed a good crop).

Also that simple decoys made of green card disrupted appropriate landings just as well as did the live ground cover.

Mechanical Pest Control

Mechanical pest control is the management and control of pests using physical means such as fences, barriers or electronic wires. It includes also weeding and change of temperature to control pests. Many farmers at the moment are trying to find sustainable ways to remove pests without harming the ecosystem.

Methods

Handpicking

The use of human hands to remove harmful insects or other toxic material is often the most common action by gardeners. It is also classified as the most direct and the quickest way to remove clearly visible pests. However, it also has equal disadvantages as it must be performed before damage to the plant has been done and before the key development of insects.

Mechanical Traps

Mechanical traps or physical attractants are used in three main ways: to efficiently trap insects, to kill them or to estimate how much many insects there are in the total landmass using sampling method. However, some traps are expensive to produce and can end up benefiting insects rather than harming them.

Differences from Integrated Pest Control

Integrated pest control refers to the use of any means to control pests once they reach unacceptable levels. Mechanical pest control is but a minor part of integrated pest control. It means only the use of physical means to control pests.

Physical Pest Control

Physical Pest Control is a method of getting rid of insects and small rodents by killing, removing, or setting up barriers that will prevent further destruction of one's plants. These methods are used primarily for crop growing, but some methods can be applied to homes as well.

Methods

Barriers

Dog control van, Rekong Peo, Himachal Pradesh, India.

Row covers are useful for keeping insects out of one's plants, typically used for horticultural crops. They are made out of either plastic or polyester. They are made thin and light to allow plants to still absorb sunshine and water from the air.

Diatomaceous earth, made from fossilized and pulverized silica shells, can be used in order to damage the protective cuticle layer of insects that have them, such as ants. When this layer is damaged, the insects become vulnerable to drying out. Unfortunately, the effectiveness of Diatomaceous earth decreases if it is wet. Therefore, it must be used often. This method was used back in the 1930s and 1940s when farmers would run dust over their fields. This would have the very same effect as diatomaceous earth.

Fire

For farmers, fire has been a powerful technique used to destroy insect breeding grounds. It is used to burn the top of the soil in order to kill the insects that lie there. Unfortunately, this can present some drawbacks. Fire can make the soil much less effective or get rid of the insects that are beneficial to the plants. Also, there is no guarantee that it will actually solve the pest problems since there may be larvae below the surface of the soil.

Firearms

Historically, firearms have been one of the primary methods used for pest control. "Garden Guns" are smooth bore shotguns specifically made to fire .22 caliber snake shot or 9mm Flobert, and are commonly used by gardeners and farmers for pest control. Garden Guns are short range weapons that can do little harm past 15 to 20 yards, and they're relatively quiet when fired with snake shot, compared to a standard ammunition. These guns are especially effective inside of barns and sheds, as the snake shot will not shoot holes in the roof or walls, or more importantly injure livestock with a ricochet. They are also used for pest control at airports, warehouses, stockyards, etc.

The most common shot cartridge is .22 Long Rifle loaded with #12 shot. At a distance of about 10 feet (3 m), which is about the maximum effective range, the pattern is about 8 inches (20 cm) in diameter from a standard rifle. Special smoothbore shotguns, such as the Marlin Model 25MG can produce effective patterns out to 15 or 20 yards using .22 WMR shotshells, which hold 1/8 oz. of #12 shot contained in a plastic capsule.

Animals

Dogs, cats, ferrets, mongoose and other animals have been historically used for pest control.

The Rat Terrier is an American dog breed with a background as a farm dog and hunting companion. Specifically bred for killing rats, today's Rat Terrier is an intelligent and active small dog that is kept both for pest control and as a family pet. Cats are also valued for companionship and for their ability to hunt vermin. Ferrets are used for hunting, or ferreting. With their long, lean build, and inquisitive nature, ferrets are very well equipped for getting down holes and chasing rodents, rabbits and moles out of their burrows. Mongooses have long been celebrated for their ability to handle venomous snakes, as immortalized in the short story *Rikki-Tikki-Tavi.*

Temperature Control

Placing produce inside of cold storage containers lengthens how long the produce lasts while also hindering the growth of insects inside of them. Another method to use is to heat, as it will kill the insect larvae in certain types of produce. An example would be with mangoes, where they are placed into a hot water bath in order to kill any eggs and larvae.

Traps

Fly paper or sticky boards are devices used in order to capture insects as they land upon the surface of the trap. They are covered in a substance that attracts insects, but are actually very sticky or poisonous. These traps are commonly used for flies or leafhoppers.

Trap strips are crops that are grown on fields with the intention of using them to attract insects and not have insects infest the other crops that are being grown. The insects can then be dealt with much more easily than if they were to have been spread throughout an entire field. Trap strips are very useful for dealing with the wheat stem sawfly. The sawflies will go only as far as they need to in order to plant their eggs. If solid stemmed plants are planted around the a crop field, then that's where the sawflies will go and the sawflies' larvae can't survive in the solid stem.

Large Scale Usage

On a much larger scale, physical control methods become much less effective because of the time that must be invested into it and because it is likely to be less economical. For example, taking care of a single tree is simple, but taking care of 500, like on a farm, would be impossible using physical control.

Chemical Methods of Pest Control

Chemical pest control methods have been used for thousands of years by civilizations which had much less knowledge than the current population. Sumerians found out that sulfur gives great results in insect extermination.

However, the actual revolution in chemical pesticides happened during the 18th and 19th century when the industrial revolution required much more efficient pest treatments in terms of scale, effectiveness and speed.

To present days, chemical pest control methods are among the major types of vermin extermination practices and despite the fact that pesticides often lead to serious health issues, chemical compounds are vastly produced and sold across the whole world.

Fungicides

Fungicides are biocidal chemical compounds or biological organisms used to kill parasitic fungi or their spores. A fungistatic inhibits their growth. Fungi can cause serious damage in agriculture, resulting in critical losses of yield, quality, and profit. Fungicides are used both in agriculture and to fight fungal infections in animals. Chemicals used to control oomycetes, which are not fungi, are also referred to as fungicides, as oomycetes use the same mechanisms as fungi to infect plants.

Fungicides can either be contact, translaminar or systemic. Contact fungicides are not taken up into the plant tissue and protect only the plant where the spray is deposited. Translaminar fungicides redistribute the fungicide from the upper, sprayed leaf surface to the lower, unsprayed surface. Systemic fungicides are taken up and redistributed through the xylem vessels. Few fungicides move to all parts of a plant. Some are locally systemic, and some move upwardly.

Most fungicides that can be bought retail are sold in a liquid form. A very common active ingredient is sulfur, present at 0.08% in weaker concentrates, and as high as 0.5% for more potent fungicides. Fungicides in powdered form are usually around 90% sulfur and are very toxic. Other active ingredients in fungicides include neem oil, rosemary oil, jojoba oil, the bacterium *Bacillus subtilis*, and the beneficial fungus *Ulocladium oudemansii*.

Fungicide residues have been found on food for human consumption, mostly from post-harvest treatments. Some fungicides are dangerous to human health, such as vinclozolin, which has now been removed from use. Ziram is also a fungicide that is toxic to humans with long-term exposure, and fatal if ingested. A number of fungicides are also used in human health care.

Natural Fungicides

Plants and other organisms have chemical defenses that give them an advantage against microorganisms such as fungi. Some of these compounds can be used as fungicides:

- Tea tree oil.
- Citronella oil.
- Jojoba oil.
- Nimbin.
- Oregano oil.
- Rosemary oil.
- Monocerin.
- Milk.

Whole live or dead organisms that are efficient at killing or inhibiting fungi can sometimes be used as fungicides:

- Bacillus subtilis.
- Ulocladium oudemansii.
- Kelp (powdered dried kelp is fed to cattle to help prevent fungal infection).
- Ampelomyces quisqualis.

Resistance

Pathogens respond to the use of fungicides by evolving resistance. In the field several mechanisms of resistance have been identified. The evolution of fungicide resistance can be gradual or sudden. In qualitative or discrete resistance, a mutation (normally to a single gene) produces a race of a fungus with a high degree of resistance. Such resistant varieties also tend to show stability, persisting after the fungicide has been removed from the market. For example, sugar beet leaf blotch remains resistant to azoles years after they were no longer used for control of the disease. This is because such mutations have a high selection pressure when the fungicide is used, but there is low selection pressure to remove them in the absence of the fungicide.

In instances where resistance occurs more gradually, a shift in sensitivity in the pathogen to the fungicide can be seen. Such resistance is polygenic – an accumulation of many mutations in different genes, each having a small additive effect. This type of resistance is known as quantitative or continuous resistance. In this kind of resistance, the pathogen population will revert to a sensitive state if the fungicide is no longer applied.

Little is known about how variations in fungicide treatment affect the selection pressure to evolve resistance to that fungicide. Evidence shows that the doses that provide the most control of the disease also provide the largest selection pressure to acquire resistance, and that lower doses decrease the selection pressure.

In some cases when a pathogen evolves resistance to one fungicide, it automatically obtains resistance to others – a phenomenon known as cross resistance. These additional fungicides are normally of the same chemical family or have the same mode of action, or can be detoxified by the same mechanism. Sometimes negative cross resistance occurs, where resistance to one chemical class of fungicides leads to an increase in sensitivity to a different chemical class of fungicides. This has been seen with carbendazim and diethofencarb.

There are also recorded incidences of the evolution of multiple drug resistance by pathogens – resistance to two chemically different fungicides by separate mutation events. For example, *Botrytis cinerea* is resistant to both azoles and dicarboximide fungicides.

There are several routes by which pathogens can evolve fungicide resistance. The most common mechanism appears to be alteration of the target site, in particular as a defence against single site of action fungicides. For example, Black Sigatoka, an economically important pathogen of banana, is resistant to the QoI fungicides, due to a single nucleotide change resulting in the replacement of one amino acid (glycine) by another (alanine) in the target protein of the QoI fungicides, cytochrome b. It is presumed that this disrupts the binding of the fungicide to the protein, rendering the fungicide ineffective. Upregulation of target genes can also render the fungicide ineffective. This is seen in DMI-resistant strains of *Venturia inaequalis*.

Resistance to fungicides can also be developed by efficient efflux of the fungicide out of the cell. *Septoria tritici* has developed multiple drug resistance using this mechanism. The pathogen had five ABC-type transporters with overlapping substrate specificities that together work to pump toxic chemicals out of the cell.

In addition to the mechanisms outlined above, fungi may also develop metabolic pathways that circumvent the target protein, or acquire enzymes that enable metabolism of the fungicide to a harmless substance.

Fungicide Resistance Management

The fungicide resistance action committee (FRAC) has several recommended practices to try to avoid the development of fungicide resistance, especially in at-risk fungicides including *Strobilurins* such as azoxystrobin.

Products should not be used in isolation, but rather as mixture, or alternate sprays, with another fungicide with a different mechanism of action. The likelihood of the pathogen's developing resistance is greatly decreased by the fact that any resistant isolates to one fungicide will be killed by the other; in other words, two mutations would be required rather than just one. The effectiveness of this technique can be demonstrated by Metalaxyl, a phenylamide fungicide. When used as the sole product in Ireland to control potato blight (*Phytophthora infestans*), resistance developed within one growing season. However, in countries like the UK where it was marketed only as a mixture, resistance problems developed more slowly.

Fungicides should be applied only when absolutely necessary, especially if they are in an at-risk group. Lowering the amount of fungicide in the environment lowers the selection pressure for resistance to develop.

Manufacturers' doses should always be followed. These doses are normally designed to give the right balance between controlling the disease and limiting the risk of resistance development. Higher doses increase the selection pressure for single-site mutations that confer resistance, as all strains but those that carry the mutation will be eliminated, and thus the resistant strain will propagate. Lower doses greatly increase the risk of polygenic resistance, as strains that are slightly less sensitive to the fungicide may survive.

It is better to use an integrative pest management approach to disease control rather than relying on fungicides alone. This involves the use of resistant varieties and hygienic practices, such as the removal of potato discard piles and stubble on which the pathogen can overwinter, greatly reducing the titre of the pathogen and thus the risk of fungicide resistance development.

Nematicides

A nematicide is a type of chemical pesticide used to kill plant-parasitic nematodes. Nematicides have tended to be broad-spectrum toxicants possessing high volatility or other properties promoting migration through the soil. Aldicarb (Temik), a carbamate insecticide marketed by Bayer CropScience, is an example of a commonly used commercial nematicide. It is important in potato production, where it has been used for control of soil-borne nematodes. Aldicarb is a cholinesterase inhibitor, which prevents the breakdown of acetylcholine in the synapse. In case of severe poisoning, the victim dies of respiratory failure. It is no longer authorised for use in the EU and, in August 2010, Bayer CropScience announced that it planned to discontinue aldicarb by 2014. Human health safety and environmental concerns have resulted in the widespread deregistration of several other agronomically important nematicides. Prior to 1985, the persistent halocarbon DBCP was a widely used nematicide and soil fumigant. However, it was banned from use after being linked to sterility among male workers; the Dow Chemical company was subsequently found liable for more than $600 million in damages.

Several natural nematicides are known. An environmentally benign garlic-derived polysulfide product is approved for use in the European Union (under Annex 1 of 91/414) and the UK as a nematicide. Another common natural nematicide is obtained from neem cake, the residue obtained after cold-pressing the fruit and kernels of the neem tree. Known by several names in the world, the tree was first cultivated in India in ancient times and is now widely distributed throughout the world. The root exudate of marigold (*Tagetes*) is also found to have nematicidal action. Nematophagous fungi, a type of carnivorous fungi, can be useful in controlling nematodes, *Paecilomyces* being one example.

Besides chemicals, soil steaming can be used in order to kill nematodes. Superheated steam is induced into the soil, which causes almost all organic material to deteriorate.

Non-fumigant Nematicides

Non-fumigant nematicides have low volatility and diffuse through the soil (generally for short distances only) dissolved in the soil solution. Their movement may be enhanced by water movement through irrigation or rainfall. If in granular formulations, their distribution may be enhanced by physical incorporation into the soil.

Side Effects of Nematicides

- Groundwater contamination with toxins.
- Exposure to chemicals.
- People who use machinery for insecticide application are at higher risk.
- Delayed harvest.
- Specific minimum time for residual effect to fade away is required which may postpone harvest.
- Pesticide poisoning.
- HIgher levels of mortality occur when certain regulations for usage are not followed.

Pesticides

A crop-duster spraying pesticide on a field.

Pesticides are substances that are meant to control pests, including weeds. The term pesticide includes all of the following: herbicide, insecticides (which may include insect growth regulators, termiticides, etc.) nematicide, molluscicide, piscicide, avicide, rodenticide, bactericide, insect repellent, animal repellent, antimicrobial, and fungicide. The most common of these are herbicides which account for approximately 80% of all pesticide use. Most pesticides are intended to serve as plant protection products (also known as crop protection products), which in general, protect plants from weeds, fungi, or insects.

A Lite-Trac four-wheeled self-propelled crop sprayer spraying pesticide on a field.

In general, a pesticide is a chemical or biological agent (such as a virus, bacterium, or fungus) that deters, incapacitates, kills, or otherwise discourages pests. Target pests can include insects, plant pathogens, weeds, molluscs, birds, mammals, fish, nematodes (roundworms), and microbes that destroy property, cause nuisance, or spread disease, or are disease vectors. Along with these benefits, pesticides also have drawbacks, such as potential toxicity to humans and other species.

Type of pesticide	Target pest group
Algicides or algaecides	Algae
Avicides	Birds
Bactericides	Bacteria
Fungicides	Fungi and oomycetes
Herbicides	Plant
Insecticides	Insects
Miticides or acaricides	Mites
Molluscicides	Snails
Nematicides	Nematodes
Rodenticides	Rodents
Slimicides	Algae, Bacteria, Fungi, and Slime molds
Virucides	Viruses

The Food and Agriculture Organization (FAO) has defined *pesticide* as:

> "Any substance or mixture of substances intended for preventing, destroying, or controlling any pest, including vectors of human or animal disease, unwanted species of plants or animals, causing harm during or otherwise interfering with the production, processing, storage, transport, or marketing of food, agricultural commodities, wood and wood products or animal feedstuffs, or substances that may be administered to animals for the control of insects, arachnids, or other pests in or on their bodies. The term includes substances intended for use as a plant growth regulator, defoliant, desiccant, or agent for thinning fruit or preventing the premature fall of fruit. Also used as substances applied to crops either before or after harvest to protect the commodity from deterioration during storage and transport".

Pesticides can be classified by target organism (e.g., herbicides, insecticides, fungicides, rodenticides, and pediculicides), chemical structure (e.g., organic, inorganic, synthetic, or biological (biopesticide), although the distinction can sometimes blur), and physical state (e.g. gaseous (fumigant)). Biopesticides include microbial pesticides and biochemical pesticides. Plant-derived pesticides, or "botanicals", have been developing quickly. These include the pyrethroids, rotenoids, nicotinoids, and a fourth group that includes strychnine and scilliroside.

Many pesticides can be grouped into chemical families. Prominent insecticide families include organochlorines, organophosphates, and carbamates. Organochlorine hydrocarbons (e.g., DDT) could be separated into dichlorodiphenylethanes, cyclodiene compounds, and other related compounds. They operate by disrupting the sodium/potassium balance of the nerve fiber, forcing the nerve to transmit continuously. Their toxicities vary greatly, but they have been phased out because of their persistence and potential to bioaccumulate. Organophosphate and carbamates largely replaced organochlorines. Both operate through inhibiting the enzyme acetylcholinesterase, allowing acetylcholine to transfer nerve impulses indefinitely and causing a variety of symptoms such as weakness or paralysis. Organophosphates are quite toxic to vertebrates and have in some cases been replaced by less toxic carbamates. Thiocarbamate and dith-iocarbamates are subclasses of carbamates. Prominent families of herbicides include phenoxy and benzoic acid herbicides (e.g. 2,4-D), triazines (e.g., atrazine), ureas (e.g., diuron), and Chloroacetanilides (e.g., alachlor). Phenoxy compounds tend to selectively kill broadleaf weeds rather than grasses. The phenoxy and benzoic acid herbicides function similar to plant growth hormones, and grow cells without normal cell division, crushing the plant's nutrient transport system. Triazines interfere with photosynthesis. Many commonly used pesticides are not included in these families, including glyphosate.

The application of pest control agents is usually carried out by dispersing the chemical in a (often hydrocarbon-based) solvent-surfactant system to give a homogeneous preparation. A virus lethality study performed in 1977 demonstrated that a particular pesticide did not increase the lethality of the virus, however combinations which included some surfactants and the solvent clearly showed that pretreatment with them markedly increased the viral lethality in the test mice.

Pesticides can be classified based upon their biological mechanism function or application method. Most pesticides work by poisoning pests. A systemic pesticide moves inside a plant following absorption by the plant. With insecticides and most fungicides, this movement is usually upward (through the xylem) and outward. Increased efficiency may be a result. Systemic insecticides, which poison pollen and nectar in the flowers, may kill bees and other needed pollinators.

In 2010, the development of a new class of fungicides called paldoxins was announced. These work by taking advantage of natural defense chemicals released by plants called phytoalexins, which fungi then detoxify using enzymes. The paldoxins inhibit the fungi's detoxification enzymes. They are believed to be safer and greener.

Uses

Pesticides are used to control organisms that are considered to be harmful. For example, they are used to kill mosquitoes that can transmit potentially deadly diseases like West Nile virus, yellow fever, and malaria. They can also kill bees, wasps or ants that can cause allergic reactions. Insecticides can protect animals from illnesses that can be caused by parasites such as fleas. Pesticides can prevent sickness in humans that could be caused by moldy food or diseased produce. Herbicides can be used to clear roadside weeds, trees, and brush. They can also kill invasive weeds that may cause environmental damage. Herbicides are commonly applied in ponds and lakes to control algae and plants such as water grasses that can interfere with activities like swimming and fishing and cause the water to look or smell unpleasant. Uncontrolled pests such as termites and mold can damage structures such as houses. Pesticides are used in grocery stores and food storage facilities to manage rodents and insects that infest food such as grain. Each use of a pesticide carries some associated risk. Proper pesticide use decreases these associated risks to a level deemed acceptable by pesticide regulatory agencies such as the United States Environmental Protection Agency (EPA) and the Pest Management Regulatory Agency (PMRA) of Canada.

DDT, sprayed on the walls of houses, is an organochlorine that has been used to fight malaria since the 1950s. Recent policy statements by the World Health Organization have given stronger support to this approach. However, DDT and other organochlorine pesticides have been banned in most countries worldwide because of their persistence in the environment and human toxicity. DDT use is not always effective, as resistance to DDT was identified in Africa as early as 1955, and by 1972 nineteen species of mosquito worldwide were resistant to DDT.

Amount Used

In 2006 and 2007, the world used approximately 2.4 megatonnes (5.3×10^9 lb) of pesticides, with herbicides constituting the biggest part of the world pesticide use at 40%, followed by insecticides (17%) and fungicides (10%). In 2006 and 2007 the U.S. used approximately 0.5 megatonnes (1.1×10^9 lb) of pesticides, accounting for 22% of the world total, including 857 million pounds (389 kt) of conventional pesticides, which are used in the agricultural sector (80% of conventional pesticide use) as well as the industrial, commercial, governmental and home & garden sectors. The state of California alone used 117 million pounds. Pesticides are also found in majority of U.S. households with 88 million out of the 121.1 million households indicating that they use some form of pesticide in 2012. As of 2007, there were more than 1,055 active ingredients registered as pesticides, which yield over 20,000 pesticide products that are marketed in the United States.

The US used some 1 kg (2.2 pounds) per hectare of arable land compared with: 4.7 kg in China, 1.3 kg in the UK, 0.1 kg in Cameroon, 5.9 kg in Japan and 2.5 kg in Italy. Insecticide use in the US has declined by more than half since 1980 (.6%/yr), mostly due to the near phase-out of organophosphates. In corn fields, the decline was even steeper, due to the switchover to transgenic Bt corn.

For the global market of crop protection products, market analysts forecast revenues of over 52 billion US$ in 2019.

Benefits

Pesticides can save farmers' money by preventing crop losses to insects and other pests; in the U.S., farmers get an estimated fourfold return on money they spend on pesticides. One study found that not using pesticides reduced crop yields by about 10%. Another study, conducted in 1999, found that a ban on pesticides in the United States may result in a rise of food prices, loss of jobs, and an increase in world hunger.

There are two levels of benefits for pesticide use, primary and secondary. Primary benefits are direct gains from the use of pesticides and secondary benefits are effects that are more long-term.

Primary Benefits

Controlling pests and plant disease vectors:

- Improved crop yields.

- Improved crop/livestock quality.

- Invasive species controlled.

Controlling human/livestock disease vectors and nuisance organisms:

- Human lives saved and disease reduced. Diseases controlled include malaria, with millions of lives having been saved or enhanced with the use of DDT alone.

- Animal lives saved and disease reduced.

Controlling organisms that harm other human activities and structures:

- Drivers view unobstructed.

- Tree/brush/leaf hazards prevented.

- Wooden structures protected.

Monetary

In one study, it was estimated that for every dollar ($1) that is spent on pesticides for crops can yield up to four dollars ($4) in crops saved. This means based that, on the amount of money spent per year on pesticides, $10 billion, there is an additional $40 billion savings in crop that would be lost due to damage by insects and weeds. In general, farmers benefit from having an increase in crop yield and from being able to grow a variety of crops throughout the year. Consumers of agricultural products also benefit from being able to afford the vast quantities of produce available year-round.

Costs

On the cost side of pesticide use there can be costs to the environment, costs to human health, as well as costs of the development and research of new pesticides.

Health Effects

Pesticides may cause acute and delayed health effects in people who are exposed. Pesticide exposure can cause a variety of adverse health effects, ranging from simple irritation of the skin and eyes to more severe effects such as affecting the nervous system, mimicking hormones causing reproductive problems, and also causing cancer. A 2007 systematic review found that "most studies on non-Hodgkin lymphoma and leukemia showed positive associations with pesticide exposure" and thus concluded that cosmetic use of pesticides should be decreased. There is substantial evidence of associations between organophosphate insecticide exposures and neurobehavioral alterations. Limited evidence also exists for other negative outcomes from pesticide exposure including neurological, birth defects, and fetal death.

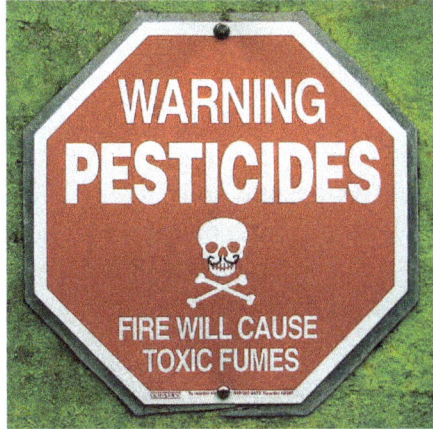

A sign warning about potential pesticide exposure.

The American Academy of Pediatrics recommends limiting exposure of children to pesticides and using safer alternatives:

Owing to inadequate regulation and safety precautions, 99% of pesticide related deaths occur in developing countries that account for only 25% of pesticide usage.

One study found pesticide self-poisoning the method of choice in one third of suicides worldwide, and recommended, among other things, more restrictions on the types of pesticides that are most harmful to humans.

A 2014 epidemiological review found associations between autism and exposure to certain pesticides, but noted that the available evidence was insufficient to conclude that the relationship was causal.

Large quantities of presumably nontoxic petroleum oil by-products are introduced into the environment as pesticide dispersal agents and emulsifiers. A 1976 study found that an increase in viral lethality with a concomitant influence on the liver and central nervous system occurs in young mice previously primed with such chemicals.

The World Health Organization and the UN Environment Programme estimate that each year, 3 million workers in agriculture in the developing world experience severe poisoning from pesticides, about 18,000 of whom die. According to one study, as many as 25 million workers in developing countries may suffer mild pesticide poisoning yearly. There are several careers aside from agriculture that may also put individuals at risk of health effects from pesticide exposure including pet groomers, groundskeepers, and fumigators.

Pesticide use is widespread in Latin America, as around US$3 billion are spent each year in the region. It has been recorded that pesticide poisonings have been increasing each year for the past two decades. It was estimated that 50–80% of the cases are unreported. It is indicated by studies that organophosphate and carbamate insecticides are the most frequent source of pesticide poisoning.

Environmental Effects

Pesticide use raises a number of environmental concerns. Over 98% of sprayed insecticides and 95% of herbicides reach a destination other than their target species, including non-target species, air, water and soil. Pesticide drift occurs when pesticides suspended in the air as particles are carried by wind to other areas, potentially contaminating them. Pesticides are one of the causes of water pollution, and some pesticides are persistent organic pollutants and contribute to soil and flower (pollen, nectar) contamination.

In addition, pesticide use reduces biodiversity, contributes to pollinator decline, destroys habitat (especially for birds), and threatens endangered species. Pests can develop a resistance to the pesticide (pesticide resistance), necessitating a new pesticide. Alternatively a greater dose of the pesticide can be used to counteract the resistance, although this will cause a worsening of the ambient pollution problem.

The Stockholm Convention on Persistent Organic Pollutants, listed 9 of the 12 most dangerous and persistent organic chemicals that were (now mostly obsolete) organochlorine pesticides. Since chlorinated hydrocarbon pesticides dissolve in fats and are not excreted, organisms tend to retain them almost indefinitely. Biological magnification is the process whereby these chlorinated hydrocarbons (pesticides) are more concentrated at each level of the food chain. Among marine animals, pesticide concentrations are higher in carnivorous fishes, and even more so in the fish-eating birds and mammals at the top of the ecological pyramid. Global distillation is the process whereby pesticides are transported from warmer to colder regions of the Earth, in particular the Poles and mountain tops. Pesticides that evaporate into the atmosphere at relatively high temperature can be carried considerable distances (thousands of kilometers) by the wind to an area of lower temperature, where they condense and are carried back to the ground in rain or snow.

In order to reduce negative impacts, it is desirable that pesticides be degradable or at least quickly deactivated in the environment. Such loss of activity or toxicity of pesticides is due to both innate chemical properties of the compounds and environmental processes or conditions. For example, the presence of halogens within a chemical structure often slows down degradation in an aerobic environment. Adsorption to soil may retard pesticide movement, but also may reduce bioavailability to microbial degraders.

Economics

Harm	Annual US cost
Public health	$1.1 billion
Pesticide resistance in pest	$1.5 billion
Crop losses caused by pesticides	$1.4 billion
Bird losses due to pesticides	$2.2 billion
Groundwater contamination	$2.0 billion
Other costs	$1.4 billion
Total costs	$9.6 billion

In one study, the human health and environmental costs due to pesticides in the United States was estimated to be $9.6 billion: offset by about $40 billion in increased agricultural production.

Additional costs include the registration process and the cost of purchasing pesticides: which are typically borne by agrichemical companies and farmers respectively. The registration process can take several years to complete (there are 70 different types of field test) and can cost $50–70 million for a single pesticide. At the beginning of the 21st century, the United States spent approximately $10 billion on pesticides annually.

Alternatives

Alternatives to pesticides are available and include methods of cultivation, use of biological pest controls (such as pheromones and microbial pesticides), genetic engineering, and methods of interfering with insect breeding. Application of composted yard waste has also been used as a way of controlling pests. These methods are becoming increasingly popular and often are safer than traditional chemical pesticides. In addition, EPA is registering reduced-risk conventional pesticides in increasing numbers.

Cultivation practices include polyculture (growing multiple types of plants), crop rotation, planting crops in areas where the pests that damage them do not live, timing planting according to when pests will be least problematic, and use of trap crops that attract pests away from the real crop. Trap crops have successfully controlled pests in some commercial agricultural systems while reducing pesticide usage; however, in many other systems, trap crops can fail to reduce pest densities at a commercial scale, even when the trap crop works in controlled experiments. In the U.S., farmers have had success controlling insects by spraying with hot water at a cost that is about the same as pesticide spraying.

Release of other organisms that fight the pest is another example of an alternative to pesticide use. These organisms can include natural predators or parasites of the pests.

Biological pesticides based on entomopathogenic fungi, bacteria and viruses cause disease in the pest species can also be used.

Interfering with insects' reproduction can be accomplished by sterilizing males of the target species and releasing them, so that they mate with females but do not produce offspring. This technique was first used on the screwworm fly in 1958 and has since been used with the medfly, the tsetse fly, and the gypsy moth. However, this can be a costly, time consuming approach that only works on some types of insects.

Push Pull Strategy

The term "push-pull" was established in 1987 as an approach for integrated pest management (IPM). This strategy uses a mixture of behavior-modifying stimuli to manipulate the distribution and abundance of insects. "Push" means the insects are repelled or deterred away from whatever resource that is being protected. "Pull" means that certain stimuli (semiochemical stimuli, pheromones, food additives, visual stimuli, genetically altered plants, etc.) are used to attract pests to trap crops where they will be killed. There are numerous different components involved in order to implement a Push-Pull Strategy in IPM.

Many case studies testing the effectiveness of the push-pull approach have been done across the world. The most successful push-pull strategy was developed in Africa for subsistence farming. Another successful case study was performed on the control of *Helicoverpa* in cotton crops in Australia. In Europe, the Middle East, and the United States, push-pull strategies were successfully used in the controlling of *Sitona lineatus* in bean fields.

Some advantages of using the push-pull method are less use of chemical or biological materials and better protection against insect habituation to this control method. Some disadvantages of the push-pull strategy is that if there is a lack of appropriate knowledge of behavioral and chemical ecology of the host-pest interactions then this method becomes unreliable. Furthermore, because the push-pull method is not a very popular method of IPM operational and registration costs are higher.

Effectiveness

Some evidence shows that alternatives to pesticides can be equally effective as the use of chemicals. For example, Sweden has halved its use of pesticides with hardly any reduction in crops. In Indonesia, farmers have reduced pesticide use on rice fields by 65% and experienced a 15% crop increase. A study of Maize fields in northern Florida found that the application of composted yard waste with high carbon to nitrogen ratio to agricultural fields was highly effective at reducing the population of plant-parasitic nematodes and increasing crop yield, with yield increases ranging from 10% to 212%; the observed effects were long-term, often not appearing until the third season of the study.

However, pesticide resistance is increasing. In the 1940s, U.S. farmers lost only 7% of their crops to pests. Since the 1980s, loss has increased to 13%, even though more pesticides are being used. Between 500 and 1,000 insect and weed species have developed pesticide resistance since 1945.

Biopesticides

Biopesticides, a contraction of 'biological pesticides', include several types of pest management intervention: through predatory, parasitic, or chemical relationships. The term has been associated historically with [biological control] – and by implication – the manipulation of living organisms. Regulatory positions can be influenced by public perceptions, thus:

- in the EU, biopesticides have been defined as "a form of pesticide based on micro-organisms or natural products".

- the US EPA states that they "include naturally occurring substances that control pests (biochemical pesticides), microorganisms that control pests (microbial pesticides), and pesticidal substances produced by plants containing added genetic material (plant-incorporated protectants) or PIPs".

They are obtained from organisms including plants, bacteria and other microbes, fungi, nematodes, *etc.* They are often important components of integrated pest management (IPM) programmes, and have received much practical attention as substitutes to synthetic chemical plant protection products (PPPs).

Types

Biopesticides can be classified into these classes:

- Microbial pesticides which consist of bacteria, entomopathogenic fungi or viruses (and sometimes includes the metabolites that bacteria or fungi produce). Entomopathogenic nematodes are also often classed as microbial pesticides, even though they are multi-cellular.

- Bio-derived chemicals. Four groups are in commercial use: pyrethrum, rotenone, neem oil, and various essential oils are naturally occurring substances that control (or monitor in the case of pheromones) pests and microbial diseases.

- Plant-incorporated protectants (PIPs) have genetic material from other species incorporated into their genetic material (*i.e.* GM crops). Their use is controversial, especially in many European countries.

- RNAi pesticides, some of which are topical and some of which are absorbed by the crop.

Biopesticides have usually no known function in photosynthesis, growth or other basic aspects of plant physiology. Instead, they are active against biological pests. Many chemical compounds have been identified that are produced by plants to protect them from pests so they are called antifeedants. These materials are biodegradable and renewable alternatives, which can be economical for practical use. Organic farming systems embraces this approach to pest control.

RNA

RNA interference is under study for possible use as a spray-on insecticide by multiple companies, including Monsanto, Syngenta, and Bayer. Such sprays do not modify the genome of the target plant. The RNA could be modified to maintain its effectiveness as target species evolve tolerance to the original. RNA is a relatively fragile molecule that generally degrades within days or weeks of application. Monsanto estimated costs to be on the order of $5/acre.

RNAi has been used to target weeds that tolerate Monsanto's Roundup herbicide. RNAi mixed with a silicone surfactant that let the RNA molecules enter air-exchange holes in the plant's surface that disrupted the gene for tolerance, affecting it long enough to let the herbicide work. This strategy would allow the continued use of glyphosate-based herbicides, but would not per se assist a herbicide rotation strategy that relied on alternating Roundup with others.

They can be made with enough precision to kill some insect species, while not harming others. Monsanto is also developing an RNA spray to kill potato beetles One challenge is to make it linger on the plant for a week, even if it's raining. The Potato beetle has become resistant to more than 60 conventional insecticides.

Monsanto lobbied the U.S. EPA to exempt RNAi pesticide products from any specific regulations (beyond those that apply to all pesticides) and be exempted from rodent toxicity, allergenicity and residual environmental testing. In 2014 an EPA advisory group found little evidence of a risk to people from eating RNA.

However, in 2012, the Australian Safe Food Foundation posited that the RNA trigger designed to change the starch content of wheat might interfere with the gene for a human liver enzyme. Supporters countered that RNA does not appear to make it past human saliva or stomach acids. The US National Honey Bee Advisory Board told EPA that using RNAi would put natural systems at "the epitome of risk". The beekeepers cautioned that pollinators could be hurt by unintended effects and that the genomes of many insects are still unknown. Other unassessed risks include ecological (given the need for sustained presence for herbicide and other applications) and the possible for RNA drift across species boundaries.

Monsanto has invested in multiple companies for their RNA expertise, including Beeologics (for RNA that kills a parasitic mite that infests hives and for manufacturing technology) and Preceres (nanoparticle lipidoid coatings) and licensed technology from Alnylam and Tekmira. In 2012 Syngenta acquired Devgen, a European RNA partner. Startup Forrest Innovations is investigating RNAi as a solution to citrus greening disease that in 2014 caused 22 percent of oranges in Florida to fall off the trees.

Examples:

Bacillus thuringiensis, a bacterial disease of Lepidoptera, Coleoptera and Diptera, is a well-known insecticide example. The toxin from *B. thuringiensis* (Bt toxin) has been incorporated directly into plants through the use of genetic engineering. The use of Bt Toxin is particularly controversial. Its manufacturers claim it has little effect on other organisms, and is more environmentally friendly than synthetic pesticides.

Other microbial control agents include products based on:

- entomopathogenic fungi (*e.g. Beauveria bassiana, Isaria fumosorosea, Lecanicillium* and *Metarhizium* spp).

- plant disease control agents: include *Trichoderma* spp. and *Ampelomyces quisqualis* (a hyper-parasite of grape powdery mildew); *Bacillus subtilis* is also used to control plant pathogens.

- beneficial nematodes attacking insect (*e.g. Steinernema feltiae*) or slug (*e.g. Phasmarhabditis hermaphrodita*) pests.

- entomopathogenic viruses (*e.g. Cydia pomonella* granulovirus).

- weeds and rodents have also been controlled with microbial agents.

Various naturally occurring materials, including fungal and plant extracts, have been described as biopesticides. Products in this category include:

- Insect pheromones and other semiochemicals.

- Fermentation products such as Spinosad (a macro-cyclic lactone).

- Chitosan: a plant in the presence of this product will naturally induce systemic resistance (ISR) to allow the plant to defend itself against disease, pathogens and pests.

- Biopesticides may include natural plant-derived products, which include alkaloids, terpenoids, phenolics and other secondary chemicals. Certain vegetable oils such as canola oil are known to have pesticidal properties. Products based on extracts of plants such as garlic have now been registered in the EU and elsewhere.

Applications

Biopesticides are biological or biologically-derived agents, that are usually applied in a manner similar to chemical pesticides, but achieve pest management in an environmentally friendly way. With all pest management products, but especially microbial agents, effective control requires appropriate formulation and application.

Biopesticides for use against crop diseases have already established themselves on a variety of crops. For example, biopesticides already play an important role in controlling downy mildew diseases. Their benefits include: a 0-Day Pre-Harvest Interval (see: maximum residue limit), the ability to use under moderate to severe disease pressure, and the ability to use as a tank mix or in a rotational program with other registered fungicides. Because some market studies estimate that as much as 20% of global fungicide sales are directed at downy mildew diseases, the integration of biofungicides into grape production has substantial benefits in terms of extending the useful life of other fungicides, especially those in the reduced-risk category.

A major growth area for biopesticides is in the area of seed treatments and soil amendments. Fungicidal and biofungicidal seed treatments are used to control soil borne fungal pathogens that cause seed rots, damping-off, root rot and seedling blights. They can also be used to control internal seed–borne fungal pathogens as well as fungal pathogens that are on the surface of the seed. Many biofungicidal products also show capacities to stimulate plant host defence and other physiological processes that can make treated crops more resistant to a variety of biotic and abiotic stresses.

Disadvantages

- High specificity: which may require an exact identification of the pest/pathogen and the use of multiple products to be used; although this can also be an advantage in that the biopesticide is less likely to harm species other than the target.

- Often slow speed of action (thus making them unsuitable if a pest outbreak is an immediate threat to a crop).

- Often variable efficacy due to the influences of various biotic and abiotic factors (since some biopesticides are living organisms, which bring about pest/pathogen control by multiplying within or nearby the target pest/pathogen).

- Living organisms evolve and increase their resistance to biological, chemical, physical or any other form of control. If the target population is not exterminated or rendered incapable of reproduction, the surviving population can acquire a tolerance of whatever pressures are brought to bear, resulting in an evolutionary arms race.

Rodenticides

Rodenticides are chemical pesticides, designed specifically for the extermination of rodents such as rats and mice.

Most rodenticides are lethal and do not serve only as repellents. They are produced and applied in the form of food which the rodents consume. It may take several hours to a few days for a rodent to be killed after consuming a rodenticide.

However, rodents often sense the threat and observe the rodenticide for a long time before consuming it. This is known as poison shyness and to reduce this, scientists now develop rodenticides with a very strong residual effect.

Instead of killing the rodent instantly, it causes dehydration and haemorrhage which cannot be stopped. This helps for avoiding problems related to rodents dying inside tiny crevices.

Common Rodenticides

Anticoagulants

An anticoagulant is the type of rodenticides that may cause injury or death to another animal after a secondary poisoning.

When consumed by rodents, anticoagulant requires one or two weeks to exterminate the rodent. The number of ingestions required depends on the generation of the anticoagulant insecticide. If they are from the first generation, multiple doses are required. In case they are a 2nd gen even a single-dose is enough to put an end to almost any rodent with a moderate size.

Anticoagulants cause death to rodents because they block the vitamin K cycle which causes a malfunction in the blood-clotting process.

A list of Anticoagulants – Rodenticides

- Warfarin.
- Chlorphacinone.
- Diphacinone.
- Bromadiolone.
- Difethialone.
- Brodifacoum.
- Bromethalin.
- Cholecalciferol.
- Zinc phosphide.
- Strychnine.

Metal Phosphides

Metal phosphides are rodenticides that kill vermin with a single dose. They are fast-acting and death from ingestion occurs in 1 to 3 days. Of all metal phosphides, zinc is the most popular for rodenticide usage, mixed with food.

If other types of rodenticides have been used before without good success, it's recommended to switch into a metal phosphide rodenticide. This is required because rats and mice give offspring too frequently and adapt to a pesticide within a short period of time, after which it has no effect on them.

A concentrate of about 0.75% to 2.0% is usually is added in rodenticides of this type. Such poison baits are distinct because of their strong garlic flavour. The odour is in fact very attractive to rodents but repels other animals which makes it very versatile as other animal species will leave it aside instead of eating it, which may disrupt the rodent extermination process.

Secondary poisoning, when metal phosphides are used, is not an issue. The risk is very low to zero because phosphides cannot accumulate in the liver of an animal similar to anticoagulants.

Hypercalcemia-causing Pesticides

Those are chemicals such as Calciferols, cholecalciferol and ergocalciferol. For people, these are healthy vitamins but to the rodent, they affect negatively the homeostasis in the body. When the rodent eat enough poisonous baits which cause hypercalcemia, the calcium levels increase so much that it starts to dissolve in the blood plasma.

The organs of the rodent that get most damaged are kidneys, stomach and lungs – the literally harden and become calcified.

Such rodenticides may not be deadly for your pet after an accidental consumption but its health may be affected. Always place poisonous baits in crevices where animals different from the targeted pest for extermination won't be able to gain access.

Additional Less familiar Rodenticides

- Barium carbonate.

- Chloralose.

- Endrin.

- Fluoroacetamide.

- Phosacetim.

- Pyrinuron.

- Scilliroside.

- Sodium fluoroacetate.

- Strychnine.

- Tetramethylenedisulfotetramine.

- Thallium sulfate.

- Nitrophenols.

Insecticides

Insecticides are substances used to kill insects. They include ovicides and larvicides used against insect eggs and larvae, respectively. Insecticides are used in agriculture, medicine, industry and by consumers. Insecticides are claimed to be a major factor behind the increase in the 20th-century's agricultural productivity. Nearly all insecticides have the potential to significantly alter ecosystems; many are toxic to humans and animals; some become concentrated as they spread along the food chain.

Insecticides can be classified into two major groups: systemic insecticides, which have residual or long term activity; and contact insecticides, which have no residual activity.

Furthermore, one can distinguish three types of insecticide. (1) Natural insecticides, such as nicotine, pyrethrum and neem extracts, made by plants as defenses against insects. (2) Inorganic insecticides, which are metals. (3) Organic insecticides, which are organic chemical compounds, mostly working by contact.

The mode of action describes how the pesticide kills or inactivates a pest. It provides another way of classifying insecticides. Mode of action is important in understanding whether an insecticide will be toxic to unrelated species, such as fish, birds and mammals.

Insecticides may be repellent or non-repellent. Social insects such as ants cannot detect non-repellents and readily crawl through them. As they return to the nest they take insecticide with them and transfer it to their nestmates. Over time, this eliminates all of the ants including the queen. This is slower than some other methods, but usually completely eradicates the ant colony.

Insecticides are distinct from non-insecticidal repellents, which repel but do not kill.

Type of Activity

Systemic insecticides become incorporated and distributed systemically throughout the whole plant. When insects feed on the plant, they ingest the insecticide. Systemic insecticides produced by transgenic plants are called plant-incorporated protectants (PIPs). For instance, a gene that codes for a specific *Bacillus thuringiensis* biocidal protein was introduced into corn (maize) and other species. The plant manufactures the protein, which kills the insect when consumed.

Contact insecticides are toxic to insects upon direct contact. These can be inorganic insecticides, which are metals and include the commonly used sulfur, and the less commonly used arsenates, copper and fluorine compounds. Contact insecticides can also be organic insecticides, i.e. organic chemical compounds, synthetically produced, and comprising the largest numbers of pesticides used today. Or they can be natural compounds like pyrethrum, neem oil etc. Contact insecticides usually have no residual activity.

Efficacy can be related to the quality of pesticide application, with small droplets, such as aerosols often improving performance.

Biological Pesticides

Many organic compounds are produced by plants for the purpose of defending the host plant from predation. A trivial case is tree rosin, which is a natural insecticide. Specifically, the production of oleoresin by conifer species is a component of the defense response against insect attack and fungal pathogen infection. Many fragrances, e.g. oil of wintergreen, are in fact antifeedants.

Four extracts of plants are in commercial use: pyrethrum, rotenone, neem oil, and various essential oils.

Other Biological Approaches

Plant-incorporated Protectants

Transgenic crops that act as insecticides began in 1996 with a genetically modified potato that produced the Cry protein, derived from the bacterium Bacillus thuringiensis, which is toxic to beetle larvae such as the Colorado potato beetle. The technique has been expanded to include the use of RNA interference RNAi that fatally silences crucial insect genes. RNAi likely evolved as a defense against viruses. Midgut cells in many larvae take up the molecules and help spread the signal. The technology can target only insects that have the silenced sequence, as was demonstrated when a particular RNAi affected only one of four fruit fly species. The technique is expected to replace many other insecticides, which are losing effectiveness due to the spread of pesticide resistance.

Enzymes

Many plants exude substances to repel insects. Premier examples are substances activated by the enzyme myrosinase. This enzyme converts glucosinolates to various compounds that are toxic to herbivorous insects. One product of this enzyme is allyl isothiocyanate, the pungent ingredient in horseradish sauces.

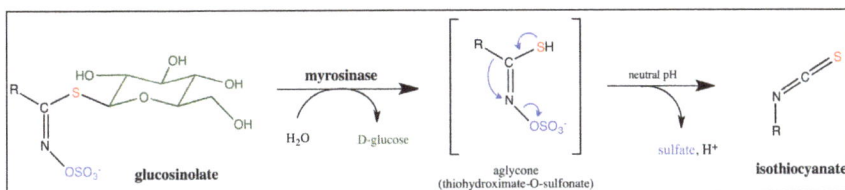

Biosynthesis of antifeedants by the action of myrosinase.

The myrosinase is released only upon crushing the flesh of horseradish. Since allyl isothiocyanate is harmful to the plant as well as the insect, it is stored in the harmless form of the glucosinolate, separate from the myrosinase enzyme.

Bacterial

Bacillus thuringiensis is a bacterial disease that affects Lepidopterans and some other insects. Toxins produced by strains of this bacterium are used as a larvicide against caterpillars, beetles, and mosquitoes. Toxins from *Saccharopolyspora spinosa* are isolated from fermentations and sold as Spinosad. Because these toxins have little effect on other organisms, they are considered more environmentally friendly than synthetic pesticides. The toxin from *B. thuringiensis* (Bt toxin) has been incorporated directly into plants through the use of genetic engineering. Other biological insecticides include products based on entomopathogenic fungi (e.g., *Beauveria bassiana*, *Metarhizium anisopliae*), nematodes (e.g., *Steinernema feltiae*) and viruses (e.g., *Cydia pomonella* granulovirus).

Synthetic Insecticide and Natural Insecticides

A major emphasis of organic chemistry is the development of chemical tools to enhance agricultural productivity. Insecticides represent a major area of emphasis. Many of the major insecticides are inspired by biological analogues. Many others are completely alien to nature.

Organochlorides

The best known organochloride, DDT, was created by Swiss scientist Paul Müller. For this discovery, he was awarded the 1948 Nobel Prize for Physiology or Medicine. DDT was introduced in 1944. It functions by opening sodium channels in the insect's nerve cells. The contemporaneous rise of the chemical industry facilitated large-scale production of DDT and related chlorinated hydrocarbons.

Organophosphates and Carbamates

Organophosphates are another large class of contact insecticides. These also target the insect's nervous system. Organophosphates interfere with the enzymes acetylcholin-esterase and other cholinesterases, disrupting nerve impulses and killing or disabling the insect. Organophosphate insecticides and chemical warfare nerve agents (such as sarin, tabun, soman, and VX) work in the same way. Organophosphates have a cumulative toxic effect to wildlife, so multiple exposures to the chemicals amplifies the toxicity. In the US, organophosphate use declined with the rise of substitutes.

Carbamate insecticides have similar mechanisms to organophosphates, but have a much shorter duration of action and are somewhat less toxic.

Pyrethroids

Pyrethroid pesticides mimic the insecticidal activity of the natural compound pyre-thrum, the biopesticide found in pyrethrins. These compounds are nonpersistent sodium channel modulators and are less toxic than organophosphates and carbamates. Compounds in this group are often applied against household pests.

Neonicotinoids

Neonicotinoids are synthetic analogues of the natural insecticide nicotine (with much lower acute mammalian toxicity and greater field persistence). These chemicals are acetylcholine receptor agonists. They are broad-spectrum systemic insecticides, with rapid action (minutes-hours). They are applied as sprays, drenches, seed and soil treatments. Treated insects exhibit leg tremors, rapid wing motion, stylet withdrawal (aphids), disoriented movement, paralysis and death. Imidacloprid may be the most common. It has recently come under scrutiny for allegedly pernicious effects on honeybees and its potential to increase the susceptibility of rice to planthopper attacks.

Ryanoids

Ryanoids are synthetic analogues with the same mode of action as ryanodine, a naturally occurring insecticide extracted from *Ryania speciosa* (Flacourtiaceae). They bind to calcium channels in cardiac and skeletal muscle, blocking nerve transmission. The first insecticide from this class to be registered was Rynaxypyr, generic name chlorant-raniliprole.

Environmental Harm

Effects on Nontarget Species

Some insecticides kill or harm other creatures in addition to those they are intended to kill. For example, birds may be poisoned when they eat food that was recently sprayed with insecticides or when they mistake an insecticide granule on the ground for food and eat it. Sprayed insecticide may drift from the area to which it is applied and into wildlife areas, especially when it is sprayed aerially.

DDT

The development of DDT was motivated by desire to replace more dangerous or less effective alternatives. DDT was introduced to replace lead and arsenic-based compounds, which were in widespread use in the early 1940s.

DDT was brought to public attention by Rachel Carson's book *Silent Spring*. One side-effect of DDT is to reduce the thickness of shells on the eggs of predatory birds. The shells sometimes become too thin to be viable, reducing bird populations. This occurs with DDT and related compounds due to the process of bioaccumulation, wherein the chemical, due to its stability and fat solubility, accumulates in organisms' fatty tissues. Also, DDT may biomagnify, which causes progressively higher concentrations in the body fat of animals farther up the food chain. The near-worldwide ban on agricultural use of DDT and related chemicals has allowed some of these birds, such as the peregrine falcon, to recover in recent years. A number of organochlorine pesticides have been banned from most uses worldwide. Globally they are controlled via the Stockholm Convention on persistent organic pollutants. These include: aldrin, chlordane, DDT, dieldrin, endrin, heptachlor, mirex and toxaphene.

Pollinator Decline

Insecticides can kill bees and may be a cause of pollinator decline, the loss of bees that pollinate plants, and colony collapse disorder (CCD), in which worker bees from a beehive or Western honey bee colony abruptly disappear. Loss of pollinators means a reduction in crop yields. Sublethal doses of insecticides (i.e. imidacloprid and other neonicotinoids) affect bee foraging behavior. However, research into the causes of CCD was inconclusive as of June 2007.

Bird Decline

Besides the effects of direct consumption of insecticides, populations of insectivorous birds decline due to the collapse of their prey populations. Spraying of especially wheat and corn in Europe is believed to have caused an 80 per cent decline in flying insects, which in turn has reduced local bird populations by a third to two thirds.

Alternatives

Instead of using chemical insecticides to avoid crop damage caused by insects, there are many alternative options available now that can protect farmers from major economic losses. Some of them are:

- Breeding crops resistant, or at least less susceptible, to pest attacks.

- Releasing predators, parasitoids, or pathogens to control pest populations as a form of biological control.

- Chemical control like releasing pheromones into the field to confuse the insects into not being able to find mates and reproduce.

- Integrated Pest Management using multiple techniques in tandem to achieve optimal results.

- Push-pull technique intercropping with a "push" crop that repels the pest, and planting a "pull" crop on the boundary that attracts and traps it.

Miticides

A miticide is a chemical pesticide that specifically targets plant mites. Mites are tiny insects that are closely related to spiders and ticks. There are several types of mites, and once established in a plant host they can damage the health of the plant. Additionally, mites can transmit viruses and diseases to your plants. Mites can be treated with both broad-spectrum and narrow-spectrum miticides.

There are several different types of mites that can affect both indoor and outdoor plants, ranging in size and color. One of the most common and easiest to recognize are spider mites. They produce fine webbing and will feed from a wide variety of plants. Other mites, such as the spruce mite, or honey locust spider mite, target specific plants. Managing these pests is crucial to preventing widespread contamination of a garden or houseplant.

Miticides often vary in the mites they target and the quickness of their kill rate. It is important for gardeners to know what type of mites their plants are infested with before selecting a particular miticide. Although mites can be treated with a broad-spectrum pesticide, these chemicals can affect beneficial insects as well as those being targeted.

Most miticides are available in spray form, and when using them it is important to focus the spray of the miticide to the undersides of plant leaves. This is particularly important to miticides that must come in direct contact with the mite in order to kill it.

Advantages and Disadvantages of Chemical Pest Control

Advantages of Chemical Pest Control

- Effectiveness.

Chemicals exterminate any pest that hasn't adapted to the deadly substance inside the agent. They eradicate fast and with an efficiency of up to 100%. Most are very easy to apply and can get in the way of pests that hide in small crevices and other hidings.

- Quickness.

Some chemicals for pest control kill slowly because of the active ingredient. However, most pesticides are designed to exterminate the vermin in less than 3-4 days, which is much faster compared to organic methods of pest control such as importation or augmentation.

- Precise targeting (localized).

Contrary to biological pest control methods, the chemical substances may target a specific area with high precision. On the other hand, if you release pest-destroying animals, there is no control over their behaviour – they may spread wherever they want.

- Easy application.

This one helps the customers a lot in their efforts towards DIY pest control with ready-to-use products. Most pesticides sold on the market are packed inside bottles, designed for easy use and application. They are readily available and spraying them on your crops takes few minutes and a little more time before that to read the instructions.

- Improving productivity.

Pesticides become more and more effective in time but sometimes at the cost of being more toxic and unsustainable. Efforts of scientists are made towards researching pesticides that cause lower pollution and side effects on human health. However, it's difficult to achieve that because animal species evolve against the chemicals and more poison is required to exterminate pests that are resistant to the old forms of pesticides.

- Sports facilities maintenance.

Sports facilities such as pitches and football terrains are endangered by pests too. If the turf is not maintained properly, it will be destroyed and the field will become unusable. Pesticides are used even there for the exterminate pests such as white grub worms, chinch bugs, bluegrass weevils, ants and more.

Disadvantages of Chemical Pest Control

- Promote evolution.

We already mentioned this but it's never useless to explain it further. The use of chemical pesticides stimulates the pest to develop resistance to the chemicals used in the pesticides. The resistance is based on alterations in the genetics of the vermin and every future generation becomes increasingly pesticide-proofed. This works very well among rodents which produce several litters per year.

- Resurgence (non-precise targeting).

Resurgence happens when the use of pesticides affects the environment and disrupt the organic pest control. The most tremendous side effects of all are when significant animal species such as bees get killed by collateral damage after spraying with chemical pesticides. Another side effect of resurgence is when pest-destroying animals such as parasitoid wasps are killed and they pray, mostly other pest insects, no longer have a natural enemy and start to multiply so quickly that completely overwhelm entire agriculture.

- Persistent organic pollutants.

POP, known as persistent organic pollutants, are extremely dangerous to the non-target organism but also affect the health of people by causing cancer, infertility or problems to the endocrine system.

Impact of Chemical Pesticides on the Environment

- Surface water contamination.

- Ground water contamination.

- Soil contamination.

- Air contamination.

- Effects on plants.

- Effects on animals.

- Direct impact on humans.

Texture of Chemical Pesticides

There are chemical pesticides in different form. You can find it in liquid or harder form and each has its specifications, advantages and disadvantages. These are the most common forms of chemical pesticides.

Granules/Pellets

The insecticides are soaked into coarse particles such as ground corn or nut shells. This way the toxic chemicals come out slowly and have a residual effect. The environment is polluted not at once but instead much slower.

Such type of insecticides is used against soil-dwelling insects for more effective results and penetration into the soil.

Dusts Chemical Pesticides

Inert particles such as ash, chalk and talk are used for making a pesticide in dust form. Their most comprehensive application is to disperse them directly onto the surface. Such a particle is usually so small and thin that they immediately stick to the body parts of insects.

This makes it easier for the active chemical ingredient to start taking effect right after the moment of application. Dust chemical pesticides are very unsuitable for application in the open as they are very easily spread by the wind and get into the human body through the respiratory system.

Soluble Powders / Wettable Powders

Those are chemical pesticides which are distributed in a powder-like form and require to mix them with water. Such foliar insecticides are described as wettable powders due to the ease of their transportation.

Emulsifiable Concentrates

Chemical pesticides based on emulsifiable concentrates do not have a residual effect on fruits and vegetable. They are mostly used in the form of sprays to treat pests in urban and industrial areas.

In standard emulsifiable pesticides, the emulsifier is usually dissolved in an organic solvent and the chemical concentration is watered with higher amounts of water.

Aerosols

Those are insecticides that have been produced and packed inside a spray can and have a solvent inside, added by the manufacturer.

References

- Mcneil, Jim (2016). "Fungi for the biological control of insect pests". Extension.org. Archived from the original on 26 May 2016. Retrieved 6 June 2016

- Pest-control, definition: maximumyield.com, Retrieved 11 March, 2019

- Gilden RC, Huffling K, Sattler B (January 2010). "Pesticides and health risks". Journal of Obstetric, Gynecologic, and Neonatal Nursing. 39 (1): 103–10. Doi:10.1111/j.1552-6909.2009.01092.x. PMID 20409108

- Chemical-pest-control-methods: pantherpestcontrol.co.uk, Retrieved 12 April, 2019

- Prabuseenivasan, Seenivasan; Jayakumar, Manickkam; Ignacimuthu, Savarimuthu (2006). "In vitro antibacterial activity of some plant essential oils". BMC Complementary and Alternative Medicine. 6: 39. Doi:10.1186/1472-6882-6-39. PMC 1693916. PMID 17134518

- Chemical-pest-control-methods: pantherpestcontrol.co.uk, Retrieved 13 May, 2019

- Eger, Christopher (28 July 2013). "Marlin 25MG Garden Gun". Marlin Firearms Forum. Outdoor Hub LLC. Retrieved 17 September 2016

- Pest-control-tools-equipment: stopthebugsmilwaukee.com, Retrieved 14 June, 2019

5
Urban Pest Control

Urban pests refer to the parasitic microorganisms which effect the human health and damage wooden support structures. It includes mosquito control, bed bugs control, rodent control, etc. This chapter closely examines these urban pest control methods to provide an extensive understanding of the subject.

Urban Integrated Pest Management

Urban integrated pest management (IPM) is concerned with a dynamic system consisting of a seminatural urban ecosystem and characteristics human components, linked by information channels. Systems science provides the perspective and organization required to deal with this complex, multiresource system.

The immediate goal of urban IPM is to provide information to environmental managers enabling them to deal with urban pest problems in ways which are ecologically sound and which satisfy the public's perceived needs (this may include altering those perceptions). The long-term objective is to arrive at new designs for urban environments that will minimize negative interactions between people and pests and minimize the use of pesticides in urban areas.

If a pesticide is used, it should be the least hazardous and target only the pest causing the problem. Taking an IPM approach is more likely to be cost effective and result in long-term pest control compared to conventional pest management practices. It is also less hazardous to human health.

Pests enter homes and buildings looking for food, water, and shelter, and there is no "one-size-fits-all" solution that will solve all pest problems. Here are some basic preventative steps that can help stop a pest problem before it starts:

- Eliminate food: Store food in hard, reusable containers with airtight, fitted lids; keep trash in a can with a tight-fitting lid and take it out on a regular basis; clean or vacuum up crumbs and spills when they occur; keep food in the refrigerator when possible.

- Eliminate water: Fix leaky and dripping pipes, faucets, and roofs; reduce humidity in basements and other moist areas of the home, such as bathrooms and kitchens; place metal screens in drains where feasible.

- Eliminate shelter: Get rid of clutter; seal cracks and crevices with silicone caulking and copper mesh where appropriate; fix or replace broken screens; maintain yard and outside areas; remove trash; prune trees, shrubs, and groundcover so they are not touching the building.

- Talk with neighbors about pest management: Pests don't stay in one location, so work together to minimize pest issues in the community.

Before you can choose and use effective IPM tactics, you must first know what pest you are dealing with. Proper identification and management may require the assistance of a licensed pest management professional. Remember, sometimes different species require different tactics (e.g., the Norway rat versus the roof rat). After the pest is identified, other sources of information can be used to determine what tactics are best to use to manage it.

An IPM toolkit is helpful to have available even if there isn't currently a pest problem. The toolkit contains basic items used when taking an IPM approach to managing pests, such as a flashlight, silicone caulk and caulk gun, copper mesh, flyswatter, cockroach bait stations, mouse snap trap, sticky traps for crawling insects, and fly paper. These items can assist with identifying the proper pest, finding out where and how the pest is getting in, and starting pest management.

Ant and Carpenter

Carpenter Ant.

About the Pest

The carpenter ant is a large, dark-colored, wingless worker ant commonly found in structural wood. It enters homes looking for food and may chew extensive tunnels in moist or rotting wood. At certain times, winged forms may be seen in great numbers.

Control Measures

Prune back nearby trees, bushes, and foliage that may be touching the home; store firewood well away from the home; eliminate excess moisture and wet wood in the house; seal cracks, crevices, and openings around pipes with silicone caulk; if ants are foraging in the house, try various ant baits. For persistent or difficult infestations, contact a licensed, reputable pest control company.

Ant, Pavement

Pavement ant.

About the Pest

Pavement ants are small and vary in color from dark brown to black. The worker ants are wingless and enter homes through cracks in search of food. They will eat many types of foodstuffs, but sweet foods are especially attractive. Ants lay down a chemical trail that other ants can use to find the food source. The colonies live outside and can be recognized by the sand piles visible in cracks of concrete, sidewalks, and at the top of foundation walls.

Control Measures

Store food in sealed containers; keep things clean and dry; caulk cracks, repair screens, and use door sweeps; don't leave pet food out or nest the bowl in a larger dish of soapy water, creating a moat; squish visible worker ants; use soapy water to wipe up the chemical trail; use ant baits and remove them when the ants are gone.

Bed Bug

Bed bug.

About the Pest

Bed bugs are chestnut brown in color and have flattened, oval bodies, which swell and turn dark red after a blood meal. They hide during the day in tiny spaces around their feeding site, such as on mattresses, furniture, and in cracks and crevices, and then feed on humans at night. Itchy welts may result from the bed bug bite. Bed bugs do not cause or spread disease, but they are difficult to control once they have entered a building or home. They can be transported from one location to another in clothing, furniture, and luggage, and they can even walk from one area to another through cracks and conduits for wiring and piping. They do not fly or jump, nor do they live on people.

Control Measures

Prevention is the best control measure avoid bringing in used furniture and mattresses; inspect baggage and clothing after traveling; toss suspect items in the dryer and dry on high for 30 minutes; seal cracks and crevices; reduce clutter. An established bed bug problem will require the assistance of a licensed, reputable pest control professional. Do not try to treat the home yourself with aerosols and "bug bombs." These products do not contact the hidden insects and may drive the bugs farther into hiding.

Centipede, House

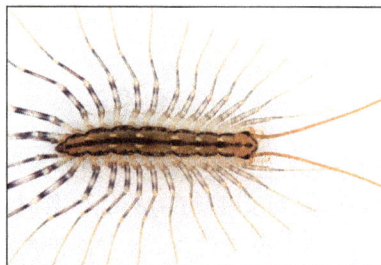

House centipede.

About the Pest

House centipedes are long, flat, and have 15 pairs of long legs as an adult (one pair per body segment). Their bodies are a brownish-yellow color and have three dark longitudinal stripes, while their legs are banded light and dark. They prefer dark, damp, cool locations and hide during the day. House centipedes can come into the house via drains and sump pumps. They can run extremely fast. They are predaceous and come out at night to feed on silverfish, cockroaches, spiders, and other small arthropods. Health concern or nuisance: Nuisance.

Control Measures

Cover drains and sump pumps with screening; remove shelter for centipedes, such as cardboard boxes; caulk and fill cracks and crevices; use a dehumidifier to remove humidity; keep things clean and dry. A high population of house centipedes may indicate a large population of prey species--monitor accordingly.

Cockroach, American

American cockroach.

About the Pest

(Often called "water bugs") The reddish-brown American cockroach is commonly found in the pantry, kitchen, bathroom, and basement. They are oval shaped with long antennae, feed on many different things, and prefer a moist, warm, dark area. During the day they tend to hide in cracks and crevices near food sources. These roaches glue their egg cases to vertical surfaces. American cockoaches tend to enter homes by crawling in from outside or up sewers and drains. These roaches can fly as well as run.

Health Concern or Nuisance

Health concern--they can carry disease-causing bacteria and can also trigger allergic and asthmatic reactions in some individuals.

Control Measures

Cover floor drains with screen and make sure there is water in all water traps; store food in sealed containers; keep things clean and dry; caulk cracks, repair screens, and use door sweeps; reduce clutter; inspect materials brought into the home; use cockroach baits; use sticky traps to monitor for new infestations; vacuum up larger infestations of roaches. Boric acid or diatomaceous earth dust can be puffed into wall voids for long-term control.

Cockroach, German

German cockroach.

About the Pest

The German cockroach is commonly found in the kitchen, bathroom, and pantry. They hide in cracks and crevices during the day. This is the most common cockroach found in homes and apartments. They are light brown or tan in color and have two dark strips running from the head to the wings. They prefer the environment to be warm and moist and will eat a large variety of foods. These cockroaches carry their egg cases around with them, protruding from their abdomens until right before the eggs hatch. Cockroaches can hitchhike from another location to your home in items such as bags and boxes.

Health Concern or Nuisance

Health concern--they can carry disease-causing bacteria and can also trigger allergic and asthmatic reactions in some individuals.

Control Measures

Store food in sealed containers; keep things clean and dry; caulk cracks; reduce clutter; inspect materials brought into the home; use cockroach baits and gels; use sticky traps to monitor for new infestations; vacuum up larger infestations of roaches. Boric acid or diatomaceous earth dust can be puffed into wall voids for long-term control.

Cockroach, Oriental

Oriental Cockroach

About the Pest

(Sometimes called "water bugs") Oriental cockroaches are very dark brown or almost black in color with a greasy-looking sheen to their bodies. These cockroaches prefer cool, dark, damp places at or below ground level, such as sewers, drains, crawl spaces, garbage cans, dumps, and trash chutes. They are not generally found crawling up walls, nor are they usually found in cupboards or on higher floors of buildings. They eat decaying organic matter both inside and outside the home.

Health Concern or Nuisance

Health concern--they can carry disease-causing bacteria and can also trigger allergic and asthmatic reactions in some individuals.

Control Measures

Check basements for sewage and drain problems; keep drains covered and basement windows and doors tight; keep areas clean and dry; caulk cracks, repair screens, and reduce clutter; store food in sealed containers. If necessary, use cockroach baits. Use sticky traps to monitor for new infestations. Boric acid or diatomaceous earth dust can be puffed into wall voids for long-term control.

Flea

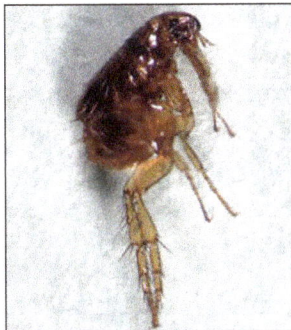

Flea.

About the Pest

Adult fleas are small, flattened, dark-colored insects with legs adapted for jumping. They do not fly. Adults spend most of their time on an animal's body. Fleas lay eggs that fall off the host animal and develop into larvae. Flea larvae, which are tiny, white, and wormlike in appearance, can live in rugs, cracks, and crevices in the floor and anywhere a pet may sleep, including outdoors during warmer weather.

Control Measures

Indoors, cover pets' sleeping areas/beds with something you can wash regularly; comb pets with flea comb; bathe pets regularly; vacuum floors, carpets, and furniture often; use borate-based carpet treatments that contain an insect growth regulator. If there are high flea populations outside, use beneficial nematodes. Discuss flea-prevention options for pets with your vet.

Fly, Fruit

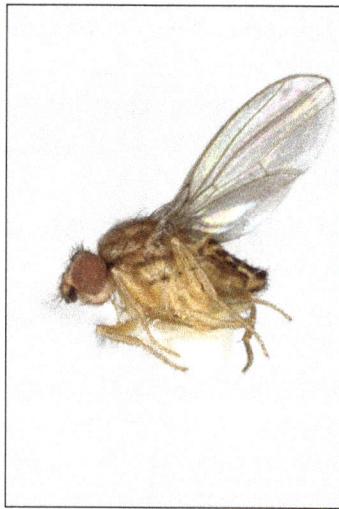

Fruit fly.

About the Pest

Fruit flies are yellowish brown to dark brown in color and usually have red eyes. They are especially attracted to ripened fruits and vegetables but can also be found in drains, garbage disposals, trash containers, and anywhere else that is moist and has fermenting material. They lay their eggs in these materials and develop very quickly from larvae to adult.

Health Concern or Nuisance

Nuisance, although they can potentially contaminate food with bacteria.

Control Measures

Eliminate sources of attraction; eat, refrigerate, or discard ripened produce; keep things clean and dry; have tight-fitting screens in windows and doors; catch flies in a trap by placing a paper funnel in a jar that has a few ounces of cider vinegar in the bottom. Clean traps weekly and replace with fresh vinegar.

Fly, House

House fly.

About the Pest

House flies have four dark stripes on the top of their middle body region. When a house fly lands on solid food, it regurgitates saliva on the food to liquefy it before ingesting it with its sponging mouthparts. They will feed on whatever food sources they find themselves on, including fecal matter and other moist, decaying matter.

Control Measures

Seal or plug cracks and gaps around windows, doors, and any pipes entering the building; repair holes in screens and make sure they fit the window tightly; keep things clean and dry; remove trash and garbage; keep garbage outside in a tightly sealed container; use a flyswatter; use sticky fly traps or tape.

Head Lice

Head louse on hair.

About the Pest

Head lice are tiny insects that spend their entire life on the human scalp. They feed on blood. Eggs (sometimes called nits) are found glued at the base of the hair shaft. Once hatched, lice nymphs and adults can walk and move around the scalp and hair as well as from one head to another if the two heads come in contact. Head lice cannot jump or fly but are transferred between people by brushes, comb, hats, or other headwear.

Control Measures

Use a metal lice comb specifically designed to remove the lice and eggs; place the comb containing lice and eggs in soapy water and flush this water down the toilet when finished with combing; boil a metal comb in water for 15 minutes before using on another person. Recheck heads every 3 days for newly hatched lice. Launder all bedding and clothing frequently in hot, soapy water and dry on hottest setting while treatment for lice is occurring in the home; place nonwashable items in tightly sealed plastic bag for two weeks or have them dry cleaned; vacuum carpets, car seats, and furniture.

Millipede

Millipede.

About the Pest

Millipedes, sometimes called "thousand-leggers," have two pairs of legs per body segment and are often dark brown in color. They have a hard outer skin and prefer to live in cool, damp places outside, such as compost or grass clipping piles, under stones and bricks, and in mulch. They are slow-moving creatures that eat decaying plant material and cannot bite people or cause damage to homes. They may accidentally enter basements, homes, and garages when plant material accumulates by doors and windows or during heavy rain or drought conditions. They only live a few days indoors.

Control Measures

Keep door sweeps and window seals in good repair; remove leaves and other debris from around doors, stairwells, and windows; sweep or vacuum up millipedes found in the home.

Moth

Indian meal moth.

About the Pest

The Indian meal moth is most commonly seen flying around in kitchens at night. It has reddish-brown markings with a copper luster on the lower two-thirds of its wings. Larvae are yellowish, greenish, or pinkish with a brown head capsule and three pairs of legs near the head. The larvae are found in common pantry items like cereal, flour, grains and grain products, and dried fruit and nuts. Larvae leave behind visible webbing as they crawl around and pupate in these products.

Control Measures

Inspect newly purchased grains before putting them in the pantry; store food in tightly sealed containers; if moths are sighted, search all dried food (e.g., cereal, pet food, oats, and other grains) and dispose of any infested food products; vacuum and thoroughly clean shelves to remove both insects and spilled food; use a flyswatter to kill adult moths. Pheromone traps containing a lure to attract male moths can be useful in detecting a problem early but will not provide control.

Moth, Webbing Clothes

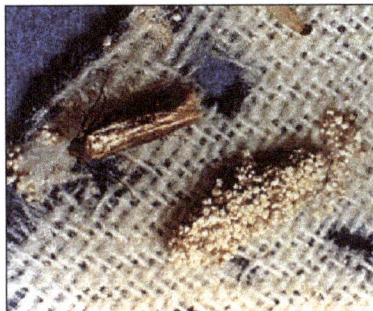

Webbing clothes moth.

About the Pest

The webbing clothes moths have tufts of red hairs on their head and golden-colored wings. The larvae have white or cream-colored bodies and a brown head. These moths feed on animal by-products like furs, hair, feathers, and wool carpets and clothes. They prefer dark environments and the adult moths will tend to run instead of fly when disturbed.

Control Measures

Launder or dry clean infested materials; hang infested rugs outside and beat or brush them; keep areas clean and well vacuumed; vacuum up the clothes moths.

Silverfish

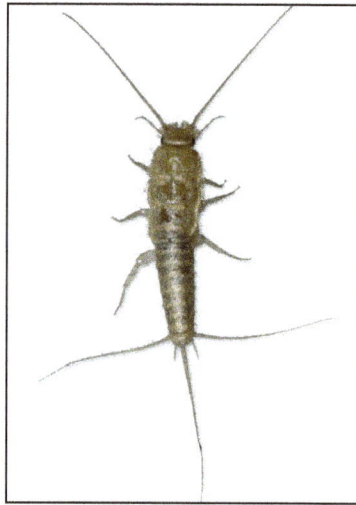

Silverfish.

About the Pest

Silverfish have silver, carrot-shaped bodies that are thicker at the head and tapered at the abdomen. They have long antennae and three long filaments coming from their abdomen. Silverfish prefer cool, moist, dark places like basements, closets, bookcases, and storage areas. They move very quickly when disturbed. Silverfish eat any substances containing starch, such as flour, cereal, books, wallpaper, and glue, but they will also feed on items high in protein.

Control Measures

Remove food sources or keep in tightly sealed container; keep things clean and dry; reduce clutter; use dehumidifier to remove humidity; seal cracks and crevices; repair leaky pipes and roof; use sticky traps.

Stink Bug, Brown Marmorated

Brown marmorated stink bug.

About the Pest

The brown marmorated stink bug has a brown, shield-shaped body, is almost as wide as it is long, and has light- brown bands on the antennae. They release an odor as a defense. These are agricultural pests; they do not harm humans, nor do they reproduce or cause damage to the home or anything inside it.

Control Measures

Seal cracks, crevices, and openings around windows, doors, pipes and chimneys; install a chimney flue cap with ½ -inch or less hardware cloth; repair damaged screens on doors and windows; remove insects from inside with a vacuum cleaner.

Termite

Termites.

About the Pest

Termites are social insects that feed on old roots, tree stumps, structural wood, wood fixtures, and paper. They are found in basement or cellar areas and in structural wood above basement walls. Wooden items buried or in contact with the ground are particularly susceptible to termite infestation. The black, winged reproductive termites are most commonly seen in infested buildings, especially between February and June when they swarm. If a piece of infested wood is opened, creamy-white worker termites may be seen. Termite damage looks like runways or passages cut into the wood. Termite tubes connecting aboveground nests with the soil may also be present.

Control Measures

Remove wooden debris from outside and around the home; replace infested wooden material; have adequate drainage in basements; do not bury scrap wood in the yard; keep fire wood piles away from the house; seal cracks and crevices in foundation; direct all surface water away from foundation. Once a home is infested, control requires the use of a licensed, reputable pest control professional.

Mosquito Control

Mosquito control manages the population of mosquitoes to reduce their damage to human health, economies, and enjoyment. Mosquito control is a vital public-health practice throughout the world and especially in the tropics because mosquitoes spread many diseases, such as malaria and the Zika virus.

Mosquito control operations are targeted against three different problems:

- Nuisance mosquitoes bother people around homes or in parks and recreational areas;

- Economically important mosquitoes reduce real estate values, adversely affect tourism and related business interests, or negatively impact livestock or poultry production;

- Public health is the focus when mosquitoes are vectors, or transmitters, of infectious disease.

Disease organisms transmitted by mosquitoes include West Nile virus, Saint Louis encephalitis virus, Eastern equine encephalomyelitis virus, Everglades virus, Highlands J virus, La Crosse Encephalitis virus in the United States; dengue fever, yellow fever, Ilheus virus, malaria, Zika virus and filariasis in the American tropics; Rift Valley fever, *Wuchereria bancrofti*, Japanese encephalitis, chikungunya and filariasis in Africa and Asia; and Murray Valley encephalitis in Australia.

Depending on the situation, source reduction, biocontrol, larviciding (killing of larvae), or adulticiding (killing of adults) may be used to manage mosquito populations. These techniques are accomplished using habitat modification, pesticide, biological-control agents, and trapping. The advantage of non-toxic methods of control is they can be used in Conservation Areas.

Monitoring Mosquito Populations

Adult mosquito populations may be monitored by landing rate counts, mechanical traps or by, lidar technology. For landing rate counts, an inspector visits a set number of sites every day, counting the number of adult female mosquitoes that land on a part of the body, such as an arm or both legs, within a given time interval. Mechanical traps use a fan to blow adult mosquitoes into a collection bag that is taken back to the laboratory for analysis of catch. The mechanical traps use visual cues (light, black/white contrasts) or chemical attractants that are normally given off by mosquito hosts (e.g., carbon dioxide, ammonia, lactic acid, octenol) to attract adult female mosquitoes. These cues are often used in combination. Entomology lidar detection has the possibility of showing the difference between male and female mosquitoes.

Monitoring larval mosquito populations involves collecting larvae from standing water with a dipper or a turkey baster. The habitat, approximate total number of larvae and pupae, and species are noted for each collection. An alternative method works by providing artificial breeding spots (ovitraps) and collecting and counting the developing larvae at fixed intervals.

Monitoring these mosquito populations is crucial to see what species are present, if mosquito numbers are rising or falling, and detecting any diseases they carry.

Source Reduction

Since many mosquitoes breed in standing water, source reduction can be as simple as emptying water from containers around the home. This is something that homeowners can accomplish. Mosquito breeding grounds can be eliminated at home by removing unused plastic pools, old tires, or buckets; by clearing clogged gutters and repairing leaks around faucets; by regularly (at most every 4 days) changing water in bird baths; and by filling or draining puddles, swampy areas, and tree stumps. Eliminating such mosquito breeding areas can be an extremely effective and permanent way to reduce mosquito populations without resorting to insecticides. However, this may not be possible in parts of the developing world where water cannot be readily replaced due to irregular water supply. Many individuals also believe mosquito control is the government's responsibility, so if these methods are not done regularly by homeowners then the effectiveness is reduced.

Open water marsh management (OWMM) involves the use of shallow ditches, to create a network of water flow within marshes and to connect the marsh to a pond or canal. The network of ditches drains the mosquito habitat and lets in fish which will feed on mosquito larvae. This reduces the need for other control methods such as pesticides. Simply giving the predators access to the mosquito larvae can result in long-term mosquito control. Open-water marsh management is used on both the eastern and western coasts of the United States.

Rotational impoundment management (RIM) involves the use of large pumps and culverts with gates to control the water level within an impounded marsh. RIM allows mosquito control to occur while still permitting the marsh to function in a state as close to its natural condition as possible. Water is pumped into the marsh in the late spring and summer to prevent the female mosquito from laying her eggs on the soil. The marsh is allowed to drain in the fall, winter, and early spring. Gates in the culverts are used to permit fish, crustaceans, and other marsh organisms to enter and exit the marsh. RIM allows the mosquito-control goals to be met while at the same time reducing the need for pesticide use within the marsh. Rotational impoundment management is used to a great extent on the east coast of Florida.

Recent studies also explore the idea of using unmanned aerial vehicles as a valid strategy to identify and prioritize water bodies where disease vectors such as *Ny. darlingi* are most likely to breed.

Nuclear Sterile Insect Technique in Mosquito Control

For the first time, a combination of the nuclear sterile insect technique (SIT) with the incompatible insect technique (IIT) was used in Mosquito Control in Guangzhou, China. The results of the recent pilot trial in Guangzhou, China, carried out with the support of the IAEA in cooperation with the Food and Agriculture Organization of the United Nations (FAO), were published in Nature on 17 July 2019.The results of this pilot trial, using SIT in combination with the IIT, demonstrate the successful near-elimination of field populations of the world's most invasive mosquito species, Aedes albopictus (Asian tiger mosquito). The two-year trial (2016-2017) covered a 32.5-hectare area on two relatively isolated islands in the Pearl River in Guangzhou. It involved the release of about 200 million irradiated mass-reared adult male mosquitoes exposed to Wolbachia bacteria.

Biocontrol

Gambusia affinis (Mosquitofish), a natural mosquito predator.

Biological control or "biocontrol" is the use of natural enemies to manage mosquito populations. There are several types of biological control including the direct introduction of parasites, pathogens and predators to target mosquitoes. Effective biocontrol agents include predatory fish that feed on mosquito larvae such as mosquitofish (*Gambusia affinis*) and some cyprinids (carps and minnows) and killifish. Tilapia also consume mosquito larvae. Direct introduction of tilapia and mosquitofish into ecosystems around the world have had disastrous consequences. However, utilizing a controlled system via aquaponics provides the mosquito control without the adverse effects to the ecosystem.

Other predators include dragonfly (fly) naiads, which consume mosquito larvae in the breeding waters, adult dragonflies, which eat adult mosquitoes, and some species of lizard and gecko. Biocontrol agents that have had lesser degrees of success include the predator mosquito *Toxorhynchites* and predator crustaceans—Mesocyclops copepods, nematodes and fungi. Predators such as birds, bats, lizards, and frogs have been used, but their effectiveness is only anecdotal.

Like all animals, mosquitoes are subject to disease. Invertebrate pathologists study these diseases in the hope that some of them can be utilized for mosquito management. Microbial pathogens of mosquitoes include viruses, bacteria, fungi, protozoa, nematodes and microsporidia.

Dead spores of the soil bacterium *Bacillus thuringiensis*, especially *Bt israelensis* (BTI) interfere with larval digestive systems. It can be dispersed by hand or dropped by helicopter in large areas. BTI loses effectiveness after the larvae turn into pupae, because they stop eating.

Two species of fungi can kill adult mosquitoes: *Metarhizium anisopliae* and *Beauveria bassiana*.

Integrated pest management (IPM) is the use of the most environmentally appropriate method or combination of methods to control pest populations. Typical mosquito-control programs using IPM first conduct surveys, in order to determine the species composition, relative abundance and seasonal distribution of adult and larval mosquitoes, and only then is a control strategy defined.

Experimental Biocontrol Methods

Introducing large numbers of sterile males is another approach to reducing mosquito numbers. This is called Sterile Insect Technique (SIT). Radiation is used to disrupt DNA in the mosquitoes and randomly create mutations. Males with mutations that disrupt their fertility are selected and released in mass into the wild population. These sterile males mate with wild type females and no offspring is produced, reducing the population size.

Another control approach under investigation for *Aedes aegypti* uses a strain that is genetically modified to require the antibiotic tetracycline to develop beyond the larval stage. Modified males develop normally in a nursery while they are supplied with this chemical and can be released into the wild. However, their subsequent offspring will lack tetracycline in the wild and never mature. Field trials were conducted in the Cayman Islands, Malaysia and Brazil to control the mosquitoes that cause dengue fever. In April 2014, Brazil's National Technical Commission for Biosecurity approved the commercial release of the modified mosquito. The FDA is the lead agency for regulating genetically-engineered mosquitoes in the United States. The review by its Center for Veterinary Medicine of a genetically engineered protein to increase the milk output of dairy cows took nine years. In the 1990s, it began a nearly 20 year review of a genetically engineered Atlantic salmon which was approved in 2015.

In 2014 and 2018 research was reported into other genetic methods including cytoplasmic incompatibility, chromosomal translocations, sex distortion and gene replacement. Although several years away from the field trial stage, if successful these other methods have the potential to be cheaper and to eradicate the Aedes aegypti mosquito more efficiently.

A pioneering experimental demonstration of the gene drive method eradicated small populations of *Anopheles gambiae*.

Trap Larva

This is a process of achieving sustainable mosquito control in an eco friendly manner by providing artificial breeding grounds with an ovitrap or an ovillanta utilizing common household utensils and destroying larvae by non-hazardous natural means such as throwing them in dry places or feeding them to larvae eating fishes like *Gambusia affinis*, or suffocating them by spreading a thin plastic sheet over the entire water surface to block atmospheric air. Shifting the water with larvae to another vessel and pouring a few drops of kerosene oil or insecticide/larvicide in it is another option for killing wrigglers, but *not preferred due to its environmental impact*. Most of the ornamental fishes eat mosquito larvae.

Trap Adult

In several experiments, researchers utilized mosquito traps. This process allowed both the opportunity to determine which mosquitoes were affected, and provided a group to be re-released with genetic modifications resulting in the OX513A variant to reduce reproduction. Adult mosquitoes are attracted inside the trap where they die of dehydration.

Oil Drip

An oil drip can or oil drip barrel was a common and nontoxic antimosquito measure. The thin layer of oil on top of the water prevents mosquito breeding in two ways: mosquito larvae in the water cannot penetrate the oil film with their breathing tube, and so drown and die; also adult mosquitoes do not lay eggs on the oiled water.

Larviciding

Control of larvae can be accomplished through use of contact poisons, growth regulators, surface films, stomach poisons (including bacterial agents), and biological agents such as fungi, nematodes, copepods, and fish. A chemical commonly used in the United States is methoprene, considered slightly toxic to larger animals, which mimics and interferes with natural growth hormones in mosquito larvae, preventing development. Methoprene is frequently distributed in time-release briquette form in breeding areas.

It is believed by some researchers that the larvae of *Anopheles gambiae* (important vectors of malaria) can survive for several days on moist mud, and that treatments should therefore include mud and soil several meters from puddles.

Adulticiding

In 1958, the National Malaria Eradication Program implemented the wide-scale use of DDT for mosquito control.

Control of adult mosquitoes is the most familiar aspect of mosquito control to most of the public. It is accomplished by ground-based applications or via aerial application of residual chemical insecticides such as Duet. Generally modern mosquito-control programs in developed countries use low-volume applications of insecticides, although some programs may still use thermal fogging. Beside fogging there are some other insect repellents for indoors and outdoors. An example of a synthetic insect repellent is DEET. A naturally occurring repellent is citronella. Indoor Residual Spraying (IRS) is another method of adulticide. Walls of properties are sprayed with an insecticide, the mosquitoes die when they land on the surface covered in insecticide.

Anti-mosquito fogging operation P.Ramachandrapuram Village in India.

To control adult mosquitoes in India, van mounted fogging machines and hand fogging machines are used.

Use of DDT

DDT was formerly used throughout the world for large area mosquito control, but it is now banned in most developed countries.

Walls on IRS-treated bathroom on the shores of Lake Victoria. The mosquitoes remain on the wall until they fall down dead on the floor.

Controversially, DDT remains in common use in many developing countries (14 countries were reported to be using it in 2009), which claim that the public-health cost of switching to other control methods would exceed the harm caused by using DDT. It is sometimes approved for use only in specific, limited circumstances where it is most effective, such as application to walls.

The role of DDT in combating mosquitoes has been the subject of considerable controversy. Although DDT has been proven to affect biodiversity and cause eggshell thinning in birds such as the bald eagle, some say that DDT is the most effective weapon in combating mosquitoes, and hence malaria. While some of this disagreement is based on differences in the extent to which disease control is valued as opposed to the value of biodiversity, there is also genuine disagreement amongst experts about the costs and benefits of using DDT.

Notwithstanding, DDT-resistant mosquitoes have started to increase in numbers, especially in tropics due to mutations, reducing the effectiveness of this chemical; these mutations can rapidly spread over vast areas if pesticides are applied indiscriminately (Chevillon *et al.* 1999). In areas where DDT resistance is encountered, malathion, propoxur or lindane is used.

Toxicant	Dosage in g/m²	Average duration of effectiveness in months
DDT	1 to 2	6 to 12
Lindane	0.5	3

Malathion	2	3
Propoxur	2	3

Mosquito Traps

A traditional approach to controlling mosquito populations is the use of ovitraps or lethal ovitraps, which provide artificial breeding spots for mosquitoes to lay their eggs. While ovitraps only trap eggs, lethal ovitraps usually contain a chemical inside the trap that is used to kill the adult mosquito and the larvae in the trap. Studies have shown that with enough of these lethal ovitraps, *Aedes* mosquito populations can be controlled. A recent approach is the automatic lethal ovitrap, which works like a traditional ovitrap but automates all steps needed to provide the breeding spots and to destroy the developing larvae.

In 2016, researchers from Laurentian University released a design for a low cost trap called an Ovillanta which consists of attractant-laced water in a section of discarded rubber tire. At regular intervals the water is run through a filter to remove any deposited eggs and larva. The water, which then contains an 'oviposition' pheromone deposited during egg-laying, is reused to attract more mosquitoes. Two studies have shown that this type of trap can attract about seven times as many mosquito eggs as a conventional ovitrap.

Some newer mosquito traps or known mosquito attractants emit a plume of carbon dioxide together with other mosquito attractants such as sugary scents, lactic acid, octenol, warmth, water vapor and sounds. By mimicking a mammal's scent and outputs, the trap draws female mosquitoes toward it, where they are typically sucked into a net or holder by an electric fan where they are collected. According to the American Mosquito Control Association, the trap will kill some mosquitoes, but their effectiveness in any particular case will depend on a number of factors such as the size and species of the mosquito population and the type and location of the breeding habitat. They are useful in specimen collection studies to determine the types of mosquitoes prevalent in an area but are typically far too inefficient to be useful in reducing mosquito populations.

Factor EOF1

Research is being conducted that indicates that dismantling a protein associated with eggshell organization, factor EOF1 (factor 1), which may be unique to mosquitoes, may be a means to hamper their reproduction effectively in the wild without creating a resistant population or affecting other animals.

Proposals to Eradicate Mosquitoes

Some biologists have proposed the deliberate extinction of certain mosquito species. Biologist Olivia Judson has advocated "specicide" of thirty mosquito species by introducing a genetic element which can insert itself into another crucial gene, to create recessive "knockout genes". She says that the *Anopheles* mosquitoes (which spread malaria) and *Aedes* mosquitoes (which spread dengue fever, yellow fever, elephantiasis, zika, and other diseases) represent only 30 out of some 3,500 mosquito species; eradicating these would save at least one million human lives per year, at a cost of reducing the genetic diversity of the family Culicidae by only 1%. She further argues that since species become extinct "all the time" the disappearance of a few more will not destroy the ecosystem: "We're not left with a wasteland every time a species vanishes. Remov-ing one species sometimes causes shifts in the populations of other species — but different need not mean worse." In addition, antimalarial and mosquito control programs offer little realistic hope to the 300 million people in developing nations who will be infected with acute illnesses this year. Although trials are ongoing, she writes that if they fail: "We should consider the ultimate swatting".

Biologist E. O. Wilson has advocated the extinction of several species of mosquito, including malaria vector Anopheles gambiae. Wilson stated, "I'm talking about a very small number of species that have co-evolved with us and are preying on humans, so it would certainly be acceptable to remove them. I believe it's just common sense".

Insect ecologist Steven Juliano has argued that "it's difficult to see what the downside would be to removal, except for collateral damage". Entomologist Joe Conlon stated that "If we eradicated them tomorrow, the ecosystems where they are active will hiccup and then get on with life. Something better or worse would take over".

However, David Quammen has pointed out that mosquitoes protect forests from human exploitation and may act as competitors for other insects. In terms of malaria control, if populations of mosquitoes were temporarily reduced to zero in a region, then this would exterminate malaria, and the mosquito population could then be allowed to rebound.

Bed Bugs Control

The term usually refers to species that prefer to feed on human blood.

Early detection and treatment are critical to successful control. According to a survey, the most commonly infested places are the mattress (98.2%), boxspring (93.6%), as well as nearby carpets and baseboards (94.1%). In fact, bed bugs thrive in areas where there is an adequate supply of available hosts, and plenty of cracks and harborages within 1.5 metres (4.9 ft) of the host.

Because treatments are required in sleeping areas and other sensitive locations, methods other than chemical pesticides are in demand. Treatments can be costly, laborious, time consuming, repetitive, and embarrassing, and may entail health risks.

—

Public Health Laws

Bed bug infestations spread easily in connecting units and have negative effects on psychological well-being and housing markets. In response, many areas have specific laws about responsibilities upon discovering a bed bug infestation, particularly in hotels and multi-family housing units, because an unprofessional level of response can have the effect of prolonging the invisible part of the infestation and spreading it to nearby units.

Common laws include responsibilities such as the following: Lessors must educate all lessees about bedbugs, lessee must immediately notify lessor in writing upon discovery of infestation, lessor must not intentionally lease infested unit, lessee must not intentionally introduce infested items, lessor must eradicate the infestation immediately every time it occurs at a professional level including all connecting units, and lessee must cooperate in the eradication process.

In a 2015 survey, reports of bed bug infestation in social media lowered the value of a hotel room to $38 for business travelers and $23 for leisure travelers.

Mapped bed bug reports graphically illustrate how difficult it can be to eliminate bed bugs in densely populated areas where many people live in adjacent units like in New York City, Los Angeles, and San Francisco.

Pesticides

Though commonly used, the pesticide approach often requires multiple visits and may not always be effective due to pesticide resistance and dispersal of the bed bugs. According to a 2005 survey, only 6.1% of companies claim to be able to eliminate bed bugs in a single visit, while 62.6% claim to be able to control a problem in 2–3 visits. Insecticide application may cause dispersal of bed bugs to neighbouring areas of a structure, spreading the infestation.

Furthermore, the problem of insecticide resistance in bed bug populations increases their opportunity to spread. Studies of bed bug populations across the United States indicate that resistance to pyrethroid insecticides, which are used in the majority of bed bugs cases, is widespread. Exterminators often require individuals to dispose of furniture and other infested materials because the pesticides are ineffective. It is advisable to break or mark these infested items to prevent their being unintentionally recycled and furthering the spread of bed bugs.

Effectiveness

The well-established resistance of bed bugs to DDT and pyrethroids has created a need for different and newer chemical approaches to the extermination of bed bugs. In 2008 a study was conducted on bed bug resistance to a variety of both old and new insecticides, with the following results, listed in order from most- to least-effective.

λ-cyhalothrin, bifenthrin, carbaryl, imidacloprid, fipronil, permethrin, diazinon, spi-nosyn, dichlorvos, chlorfenapyr, and DDT. Note that the first of these, λ-cyhalothrin, is itself a pyrethroid-based insecticide— in the past it has been used principally for the treatment of cotton crops and so bed bugs have not yet developed a genetic resistance to it.

Up until the 1990s chlorpyrifos was used as an agent with longterm effect, but the EC biocide declaration *98/8* prohibited the use from August 2008 onward.

Some manufacturers also offer fumigants containing sulfuryl fluoride.

Disadvantages

Non-residue methods of treatment such as steaming and vacuuming are preferable to the contamination of mattresses, pillows and bed covers with insecticides. The possible health effects of pesticides on people and pets ranging from allergic reactions to cancer and acute neurotoxicity have to be considered, as well as the dispersal of bed bugs to neighbouring dwellings due to repellent effects of insecticides.

Bedbugs prefer to hide in and around the bed frame but it can still be a good idea to put a tight cotton cover on mattress and bedding to prevent access.

Pesticide Resistance

Bed bugs are largely resistant to various pesticides including DDT and organophos-phates. Most populations have developed a resistance to pyrethroid insecticides. Al-though now often ineffective, the resistance to pyrethroid allows for new chemicals that work in different ways to be investigated, so chemical management can continue to be one part in the resolving of bed bug infestations. There is growing interest in both syn-thetic pyrethroid and the pyrrole insecticide, chlorfenapyr. Insect growth regulators, such as hydroprene (Gentrol), are also sometimes used.

Populations in Arkansas have been found to be highly resistant to DDT, with an LD_{50} of more than 100,000 ppm. DDT was seen to make bed bugs more active in studies conducted in Africa.

Bed bug pesticide-resistance appears to be increasing dramatically. Bed bug popu-lations sampled across the U.S. showed a tolerance for pyrethroids several thousand times greater than laboratory bed bugs. New York City bed bugs have been found to be 264 times more resistant to deltamethrin than Florida bed bugs due to mutations and evolution. Products developed in the mid 2010s combine neonicotinoids with py-rethroids, but according to a January 2016 survey published by the Journal of Medical Entomology, bed bug resistance in two major US cities now includes neonicotinoids.

A population genetics study of bed bugs in the United States, Canada, and Australia using a mitochondrial DNA marker found high levels of genetic variation. This suggests the studied bed bug populations did not undergo a genetic bottleneck as one would expect from insecticide control during the 1940s and 1950s, but instead, that populations may have been maintained on other hosts such as birds and bats. In contrast to the high amount of genetic variation observed with the mitochondrial DNA marker, no genetic variation in a nuclear RNA marker was observed. This suggests increased gene flow of previously isolated bed bug populations, and given the absence of barriers to gene flow, the spread of insecticide resistance may be rapid.

Physical Isolation

Isolation of humans is attempted with numerous devices and methods including zippered bed bug-proof mattress covers, bed-leg moat devices, and other barriers. However, even with isolated beds, bed bug infestations persist if the bed itself is not free of bed bugs, or if it is re-infested, which could happen quite easily.

It is convenient to place medium-sized belongings in sealed transparent plastic bags (such as plastic bags for freezing; larger models exist as well). Once closed, the tightness should be verified by pressing the bag and ensuring that air doesn't exit. It is as well convenient to mark these sealed bags as 'contaminated'/'decontaminated'.

Inorganic Materials

A sample of food-grade diatomaceous earth.

Inorganic materials such as diatomaceous earth or amorphous silica gel may be used in conjunction with other methods to manage a bed bug infestation, provided they are used in a dry environment. Upon contact with such dust-like materials, the waxy outer layer of the insect's exoskeleton is disrupted, which causes them to dehydrate.

Food-grade diatomaceous earth has been widely used to combat infestations. However, it can take weeks to have a significant effect. Studies examined and compared diatomaceous earth and synthetically produced, pure amorphous (i.e. non-crystalline) silica, so-called silica gel. They investigated the use of these substances as a stand-alone treatment in real-life scenarios, and compared them to usual poisonous agents. They found that the effect of diatomaceous earth was surprisingly low when used in real-life scenarios, while the synthetic product was extremely effective and fast in killing bed bugs in such settings.

Silica gel was also more effective than usual poisonous pesticides (particularly in cases with pesticide resistant bugs). When applied after being mixed with water and then sprayed, the outcome for silica gel was significantly lower, but still distinctly better than for the natural silica (used dry). Authors argued that the reason for the poor outcome for diatomaceous earth as a stand-alone treatment was multi-factorial. When tested in laboratory where the bed bugs had intensive, prolonged contact with diatomaceous earth and no access to a host, diatomaceous earth performed very well. Silica gel, on the other side, performed *in vitro* consistently well even if applied to bed bugs in extremely low doses and with very slight and short (often only seconds or few minutes) contact to the substance.

Although occasionally applied as a safe indoor pesticide treatment for other insects, boric acid is ineffectual against bed bugs because bed bugs do not groom.

Organic Materials

Bean Leaves

A traditional Balkan method of trapping bed bugs is to spread bean leaves in infested areas. The trichomes (microscopic hooked hairs) on the leaves trap the bugs by piercing the tarsi joints of the bed bug's arthropod legs. As a bug struggles to get free, it impales itself further on the bean leaf's trichomes. The bed bugs and leaves then can be collected and destroyed. Researchers are examining ways to reproduce this capability with artificial materials.

Essential oils

Many claims have been made about essential oils killing bed bugs. However, they are unproven. The FTC is now filing a suit against companies making these claims about these oils, specifically about cedar, cinnamon, lemongrass, peppermint, and clove oils.

Contaminated Belongings

Disposal of items from the contaminated area can reduce the population of bed bugs and unhatched eggs. Removal of items such as mattresses, box springs, couches etc. is costly and usually insufficient to eradicate infestation because of eggs and adults hiding in surrounding areas. If the entire infestation is not eliminated prior to bringing new or cleaned personal and household items back into a home, these items will likely become infested and require additional treatment.

Treating clothing, shoes, linens, and other household items within the affected environment is difficult and frequently ineffective because of the difficulty of keeping cleaned items quarantined from infestation. Many bed bug exterminating specialists recommend removing personal and household items from the infested structure. Many metropolitan areas offer more effective treatments such as high-heat dryers and dry cleaning with PERC with the added benefit of the treated items remaining stored until the affected home's bed bug infestation is eradicated.

The improper disposal of infested furniture also facilitates the spread of bed bugs. Marking the discarded items as infested can help prevent infesting new areas. Bed bugs can go without feeding for 20 to 400 days, depending on temperature and humidity. Older stages of nymphs can survive longer without feeding than younger ones, and adults have survived without food for more than 400 days in the laboratory at low temperatures. Adults may live up to one year or more, and there can be up to four successive generations per year.

Vacuuming

Vacuuming helps with reducing bed bug infestations, but does not eliminate bed bugs hidden inside of materials. Also, unless the contents of the vacuum are emptied immediately after each use, bedbugs may crawl out through the vacuum's hoses and re-establish themselves. Vacuuming with a large bristle attachment can also aid in removing hidden bugs as well.

Heat Treatment

Steam

Steam treatment can effectively kill all stages of bed bugs. To be effective, steam treatment must reach 150–170 degrees Fahrenheit (65 - 75 degrees C) for a sustained period. Unfortunately, bed bugs hide in a diversity of places, making steam treatment very tedious, labour-intensive and time consuming. There is also the risk of the steam not penetrating materials enough to kill hidden bed bugs. The steam may also damage materials such as varnished wood, or cause mold from the moisture left behind. Effective treatment requires repeated and very thorough steaming of the mattress, box spring, bed frame, bed covers, pillows, not to mention other materials and objects within the infested room, such as carpets and curtains.

Infested clothes can be effectively treated by a high-temperature ironing with vapor. If performed meticulously, this method yields faster disinfection compared to high-temperature washing in a washing machine. However, attention should be paid in order to avoid bedbug escape from the ironed clothes.

For volumetric objects (e.g. pillow, blanket, sleeping bag, rug), boiling in a large saucepan for more than 10 minutes represents a reliable method. In this manner, the lethal temperatures propagate with certainty deep inside the object, which is not necessarily the case of a washing machine cleaning cycle.

For smaller objects, pouring boiling water from a kettle onto the object located in a basin may be enough to kill bed bugs and eggs.

Clothes Dryers

Clothes dryers can be used for killing bed bugs in clothing and blankets. Infested clothes and bedding are first washed in hot water with laundry detergent then placed in the dryer, and then after the items are completely dry, continue drying for at least 20 minutes longer at high heat. However, this does not eliminate bed bugs in the mattress, bed frame and surrounding environment. Sterilized fabrics from the dryer are thus easily re-infested. Continually treating materials in this fashion is labour-intensive, and in itself does not eliminate the infestation.

Hot Boxes

Placing belongings in a hot box, a device that provides sustained heat at temperatures that kills bedbugs, larvae, and eggs, but that does not damage clothing, is an option. Pest control companies often rent the devices at nominal cost and it may make sense for frequent travelers to invest in one.

Building Heat Treatment

This method of bed bug control involves raising room temperatures to or above the killing temperature for bed bugs, which is around 45 °C (113 °F). Heat treatments are generally carried out by professionals, and may be performed in a single area or an entire building. Heat treatment is generally considered to be the best method of eradication because it is capable of destroying an entire infestation with a single treatment.

HEPA air filtration is normally used during any heat treatment to capture particulate and biological matter that may be aerosolized during the heating process.

Freezing

Bed bugs can be killed by a direct one-hour exposure to temperatures of −16 °C (3 °F), however, bed bugs have the capacity for rapid cold hardening, i.e. an hour-long exposure to 0 °C (32 °F) improved their subsequent tolerance of −14 to −16 °C (7 to 3 °F), so this may need to be maintained for longer. Freezer temperatures at or below −16 °C (3 °F) should be sufficient to eliminate bed bugs and can be used to decontaminate household objects. This temperature range should be effective at killing eggs as well as all stages of bugs. Higher temperatures however are not effective, and survival is estimated for temperatures above −12 °C (10 °F) even after 1 week of continuous exposure.

This method requires a freezer capable of maintaining, and set to, a temperature below −16 °C (3 °F). Most home freezers are capable of maintaining this temperature.

Fungus

Preliminary research has shown the fungus *Beauveria bassiana*, which has been used for years as an outdoor organic pesticide, is also highly effective at eliminating bed bugs exposed to cotton fabric sprayed with fungus spores. It is also effective against bed bug colonies due to the spores carried by infected bugs back to their harborages. Unlike typical insecticides, exposure to the fungus does not kill instantly, but kills bugs within five days of exposure. Some people, especially those with compromised immune systems, may react negatively to the concentrated presence of the fungus directly following an application.

Drugs

Early research shows that the common drug taken to get rid of parasitic worms, ivermectin (Stromectol), also kills bed bugs when taken by humans at normal doses. The drug enters the human bloodstream and if the bedbugs bite during that time, the bedbug will die in a few days. Ivermectin is also effective against mosquitoes, which can be useful controlling malaria.

Predators

Natural enemies of bed bugs include the masked hunter (or masked bed bug hunter) insect, cockroaches, ants, spiders (particularly *Thanatus flavidus*), mites, and centipedes (particularly the house centipede *Scutigera coleoptrata*). However, biological pest control is not considered practical for eliminating bed bugs from human dwellings.

Rodent Control

Rodents are known to be carriers of diseases and can cause damage to homes. This is achieved by eliminating any food sources, sealing even the smallest entries into homes, and successfully trapping rodents.

You play a key role in rodent control and maintaining a clean community is the best prevention tool in rodent control.

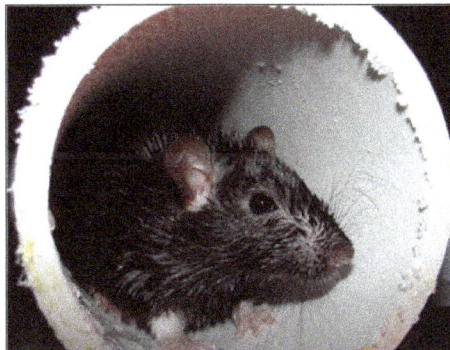

1. Clean Up: Clean up by eliminating rodent food sources and nesting sites.

- Clean up your home, pick up or eliminate clutter inside and outside your home.

- Store food in thick plastic or metal containers with tight fitting lids.

- Put food away after use. Clean up spilled food right away. Wash dishes and cooking utensils after use.

- Store all garbage and food waste in thick plastic or metal cans with tight fitting lids. Dispose of garbage on a regular basis.

- Outside your home: Get rid of old vehicles, tires, and clutter that mice and rats might use as homes.

- Keep outside cooking areas and grills clean.

- Do not feed wildlife; do not leave pet food or water bowls outside in feeding dishes.

- Keep bird feeders away from your home; utilize squirrel guards on all bird feeders.

- Keep compost bins away from your home.

- Keep woodpiles away from your home. Elevate wood piles at least eight inches off ground.

- Keep grass, shrubbery, and vegetation cut short and well-trimmed to eliminate possible nesting site outside your home.

Keep food in thick plastic or metal containers with tight lids.

Keep grass, shrubbery, and vegetation cut short.

2. Seal Up: Seal up all small holes and entries into your home. Mice can squeeze through a hole the size of a nickel and rats can squeeze through a hole the size of a half dollar.

- Locate and seal all holes, gaps, and openings in your home ¼ inch or larger.

- Where to look inside your home: inside, under, and behind kitchen cabinets, refrigerators and stoves. Look inside closets, near floor corners, around the fireplace, around door frames, around the pipes under sinks and washing machines, around floor vents and dryer vents, inside attic, and in the basement or crawl space.

- Where to look outside your home: in the roof among the rafters, gables, and eaves, around windows, around doors, around the foundations, attic vents and crawl spaces vents, under doors, around holes for electrical, plumbing, cable, and gas lines.

- Seal holes with steel wool. Put caulk around the steel wool to keep it place. Use lath screen, lath metal, cement, hardware cloth, or metal sheeting to fix large holes. These materials can be found at your local hardware store.

- Use flashing around base of house and doors.

- Shed, garages, and outside buildings should also be sealed to prevent the entrance of rodents.

- If you do not remember to seal up entry holes in your home, rodents will continue to get inside.

Seal holes on exterior of your house.

3. Trap Up: Trap up rodents in your home to successfully reduce the rodent population.

- Choose appropriate trap and carefully read instructions before setting trap.

- Place traps in rodent runways along walls between nests and feeding areas.

- Place traps in attics, basements, and crawlspaces and other areas that do not have regular human traffic, set traps in areas where there is evidence of frequent rodent activity.

- Contact a licensed Pest Control Operator to treat burrows.

Place snap traps inside your home.

Rat burrows: Most rats live in nests or burrows. Burrows are holes in dirt, wood or concrete that range from 1-inch to 4-inches wide. Rodents cannot be eliminated by blocking their burrows, they will dig another burrow.

If you find a rat burrow on your property contact a licensed pest control agent to treat the burrows.

Rat chew in walls.

Active rat burrow.

Diseases Spread by Rodents

There are 35 diseases that can be traces to rodents. Some from handling rodents, contact with rodent feces, urine, or saliva, or through rodent bites. Some diseases know to be spread through direct contact with rodents include Hantavirus, Pulmonary Syndrome, or Tularemia. Other diseases are spread from indirect contact such as Babesiosis or Lyme Disease which can be spread by ticks on the rodents. It is recommended that you avoid all contact with rodents, rodents droppings, and urine as these can be sources of diseases.

References

- Cohrssen, John J.; Miller, Henry I. (8 July 2014). "How Many Regulators Does It Take to Kill A Mosquito?". Forbes. Retrieved 10 October 2014

- Common-urban-pests-identification-prevention-and-control: extension.psu.edu, Retrieved 15 July, 2019

- Tabachnick WJ (September 2016). "Research Contributing to Improvements in Controlling Florida's Mosquitoes and Mosquito-borne Diseases". Insects. 7 (4): 50. Doi:10.3390/insects7040050. PMC 5198198. PMID 27690112

- Control-of-Rodents-in-an-Urban-Area-Flyer-, View, documentcenter: watertown-ma.gov, Retrieved 16 August, 2019

- Miller, Dini (2008). "Bed bugs (hemiptera: cimicidae: Cimex spp.)". In Capinera, John L. (ed.). Encyclopedia of Entomology (Second ed.). Springer. P. 414. ISBN 9781402062421

Permissions

All chapters in this book are published with permission under the Creative Commons Attribution Share Alike License or equivalent. Every chapter published in this book has been scrutinized by our experts. Their significance has been extensively debated. The topics covered herein carry significant information for a comprehensive understanding. They may even be implemented as practical applications or may be referred to as a beginning point for further studies.

We would like to thank the editorial team for lending their expertise to make the book truly unique. They have played a crucial role in the development of this book. Without their invaluable contributions this book wouldn't have been possible. They have made vital efforts to compile up to date information on the varied aspects of this subject to make this book a valuable addition to the collection of many professionals and students.

This book was conceptualized with the vision of imparting up-to-date and integrated information in this field. To ensure the same, a matchless editorial board was set up. Every individual on the board went through rigorous rounds of assessment to prove their worth. After which they invested a large part of their time researching and compiling the most relevant data for our readers.

The editorial board has been involved in producing this book since its inception. They have spent rigorous hours researching and exploring the diverse topics which have resulted in the successful publishing of this book. They have passed on their knowledge of decades through this book. To expedite this challenging task, the publisher supported the team at every step. A small team of assistant editors was also appointed to further simplify the editing procedure and attain best results for the readers.

Apart from the editorial board, the designing team has also invested a significant amount of their time in understanding the subject and creating the most relevant covers. They scrutinized every image to scout for the most suitable representation of the subject and create an appropriate cover for the book.

The publishing team has been an ardent support to the editorial, designing and production team. Their endless efforts to recruit the best for this project, has resulted in the accomplishment of this book. They are a veteran in the field of academics and their pool of knowledge is as vast as their experience in printing. Their expertise and guidance has proved useful at every step. Their uncompromising quality standards have made this book an exceptional effort. Their encouragement from time to time has been an inspiration for everyone.

The publisher and the editorial board hope that this book will prove to be a valuable piece of knowledge for students, practitioners and scholars across the globe.

Index

www.ingramcontent.com/pod-product-compliance
Lightning Source LLC
Chambersburg PA
CBHW061955190326
41458CB00009B/2877

* 9 7 8 1 6 4 1 1 6 5 2 1 1 *